Learn to Use Microsoft Word 2016

Michelle N. Halsey, PMP, CSM

ISBN-10: 1-64004-254-7

ISBN-13: 978-1-64004-254-4

Silver City Publications & Training, L.L.C.
P.O. Box 1914
Nampa, ID 83653
https://www.silvercitypublications.com/shop/

Contents

Chapter 1 – Opening Word

In this chapter, you will learn how to open Word, where you will first encounter the Recent list and other ways you can start a document. You will learn how to open files and how to create a blank document or a document from a template.

Opening Word

To open Word in Windows 8, use the following procedure.

Step 1: From the Start page, select the Word 2016 icon.

Use this procedure if using Windows 7or previous versions of Widows:

Step 1: Select the Start icon from the lower left side of the screen.

Step 2: Select All Programs.

Step 3: Select Microsoft Office.

Step 4: Select Microsoft Office Word 2016.

Using the Recent List

To open a document from the Recent list, use the following procedure.

Step 1: Select the document that you want to open from the Recent list.

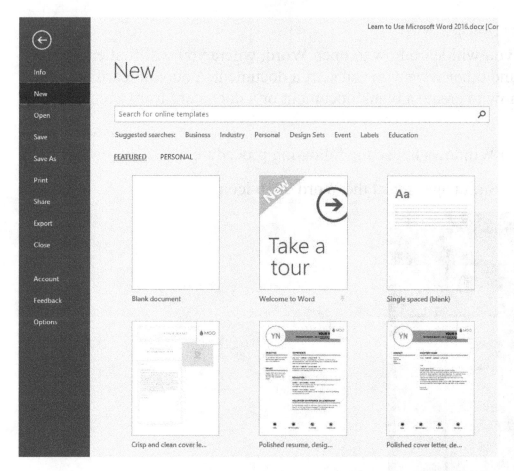

To pin an item on the Recent list, use the following procedure.

Step 1: Click the pin on the right side of the Recent list item.

The item moves to the top section of the Recent list.

To unpin an item, click the pin on the right side of the Recent list again. The item returns to the previous location in the Recent list.

Opening Files

To open a document, use the following procedure.

Step 1: Select Open Other Documents from the bottom of the Recent list. Or select Open from the Backstage View.

Step 2: Select one of the Places you would like to look for the document. The default options are Recent Documents, your Microsoft OneDrive location, and your Computer.

Step 3: To open a document from the OneDrive or your computer, select Browse.

Step 4: In the Open dialog box, navigate to the location of the file you want to open. Select it and select Open.

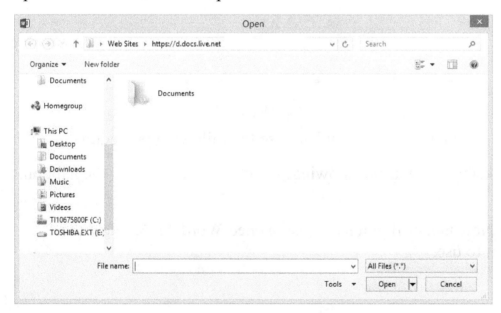

Creating a Blank Document

To create a blank document, use the following procedure.

Step 1: If the Backstage view is not showing, select the File tab from the Ribbon. Select New.

Step 2: From the New tab, or if you have just opened Word 2016, select Blank Document.

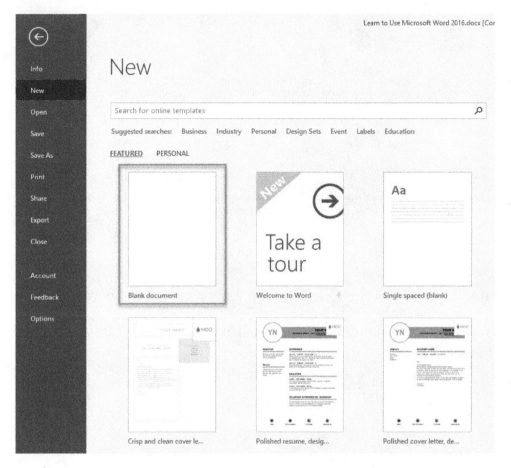

Creating a Document from a Template

To create a blank document from a template, use the following procedure.

Step 1: If the Backstage view is not showing, select the File tab from the Ribbon. Select New.

Step 2: From the New tab, or if you have just opened Word 2016, select the template you want to use.

Step 3: Select Create.

You can use the left and right arrows to review the other templates in the current search.

To search for a template and filter the results, use the following procedure.

Step 1: Select one of the Suggested Search terms or enter a term in the Search box and press Enter.

Step 2: To apply a filter, select the Filter term from the list on the right side of the screen.

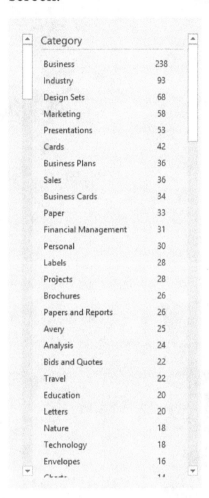

Category	
Business	238
Industry	93
Design Sets	68
Marketing	58
Presentations	53
Cards	42
Business Plans	36
Sales	36
Business Cards	34
Paper	33
Financial Management	31
Personal	30
Labels	28
Projects	28
Brochures	26
Papers and Reports	26
Avery	25
Analysis	24
Bids and Quotes	22
Travel	22
Education	20
Letters	20
Nature	18
Technology	18
Envelopes	16

Step 3: To return to the list of templates, select Home.

Chapter 2 – Working with the Interface

In this chapter, we will introduce you to the Word 2016 interface, which uses the Ribbon from the previous two versions of Word. You will get a closer look at the Ribbon, as well as the Navigation pane and the Status bar. You will also learn how to manage your Microsoft account right from a new item above the Ribbon. This chapter introduces you to the Backstage view, where all the functions related to your files live. You will learn how to save files. Finally, we will look at closing files and closing the application.

Understanding the Interface

The Word interface includes the Ribbon, the Navigation pane, the document window, the Quick Access toolbar, and the Status bar.

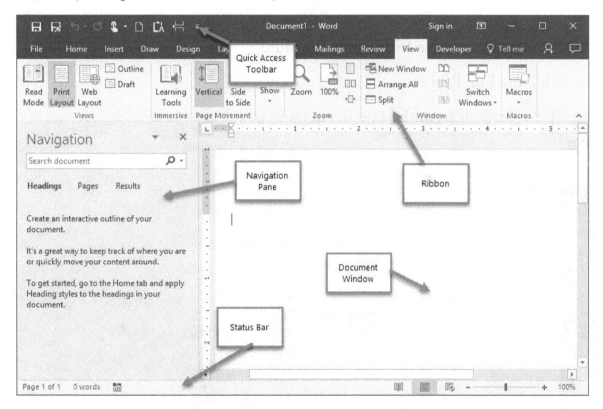

Each Tab in the Ribbon contains many tools for working with your document. To display a different set of commands, click the Tab name. Buttons are organized into groups according to their function.

The Quick Access toolbar appears at the top of the Word window. It provides you with one-click shortcuts to commonly used functions, like save, undo, and redo.

The Navigation pane allows you to quickly move through headings, pages, or search results.

The Status bar shows your current page, the word count, the language setting for proofing, and if any macros are currently running. It also allows you to quickly change your view or zoom of the document.

To zoom in or out, use the following procedure.

Step 1: Click the minus sign in the Status bar to zoom out. Click the plus sign in the Status bar to zoom in. You can also drag the slider to adjust the zoom.

You can also click the number percentage to open the Zoom dialog box.

About Your Account and Feedback

The account options use the following procedure.

Step 1: Click the File Ribbon.

Step 2: Click Account from the File menu options.

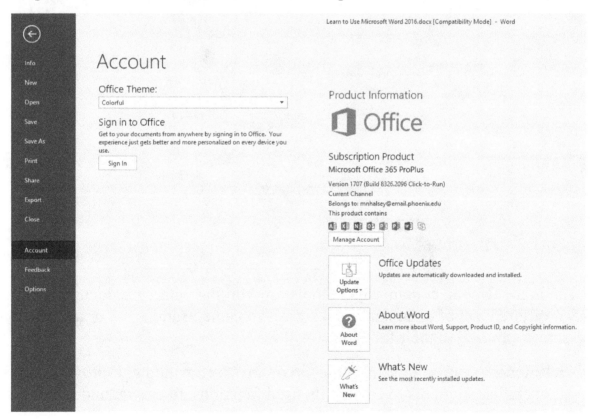

Step 3: Click Sign In or Manage Account to make account adjustments.

To send feedback to Microsoft, use the following procedure.

Step 1: Select the File Ribbon.

Step 2: Select the Feedback option.

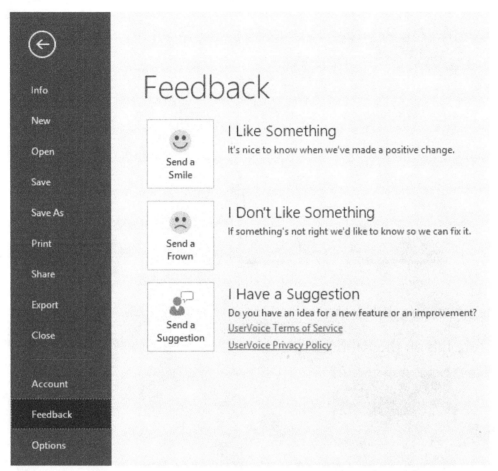

Step 3: Select the desired Feedback option.

Step 4: Enter the information requested in the Microsoft Office Feedback dialog. Select Send.

An Introduction to Backstage View

The Backstage View, use the following procedure.

Step 1: Select the File tab on the Ribbon.

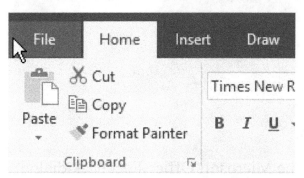

Word displays the Backstage View, open to the Info tab by default. A sample is illustrated below.

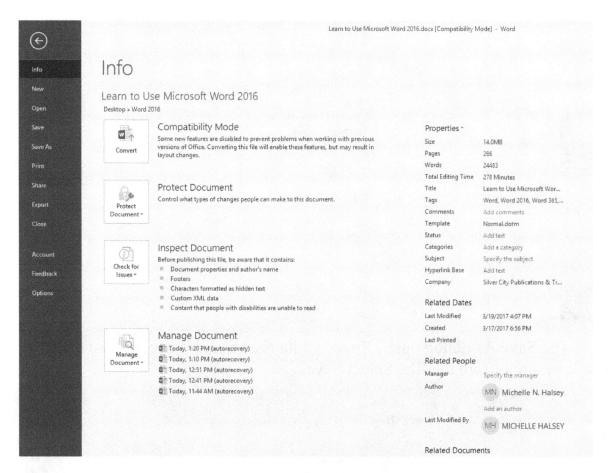

Saving Files

To save a document that has not been previously saved, use the following procedure.

Step 1: Select the File tab on the Ribbon.

Step 2: Select the Save As command in the Backstage View.

Step 3: Select the Place where you want to save the document.

Step 4: If you choose your OneDrive, you can select the Documents folder. If you choose your Computer, select your Current Folder or one of your Recent Folders. Or in either place, you can choose Browse to select a new location.

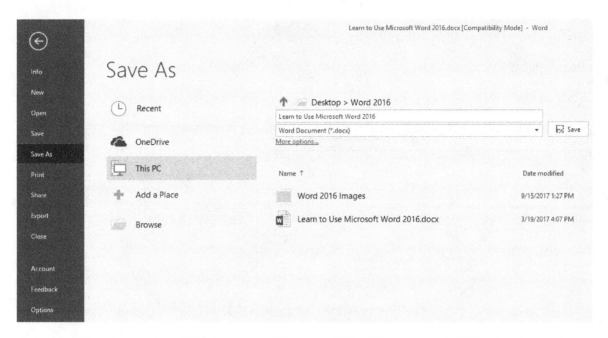

Step 5: The Save As dialog opens. Enter a File Name, and if desired, navigate to a new location to store the file. Select Save.

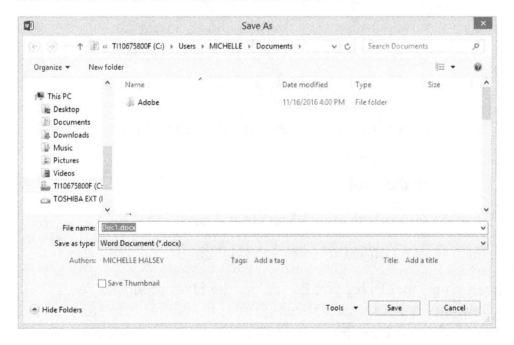

To close a file, use the following procedure.

Step 1: Select the File tab from the Ribbon.

Step 2: Select Close from The Backstage View.

If you have not saved your file, you will see the following message.

To close the application (if only one document is open), use the following procedure.

Step 1: Click the X at the top right corner of the window.

Chapter 3 – Your First Document

In this chapter, you will create your first document. You will learn how to type text and select it with a mouse or keyboard. With text selected, you can edit or delete text, or you can use the mouse to drag and drop text to a new location. You will also learn how to insert symbols and numbers. Finally, this chapter will cover how to start a new page.

Typing Text

The following diagram shows the cursor location in a blank document.

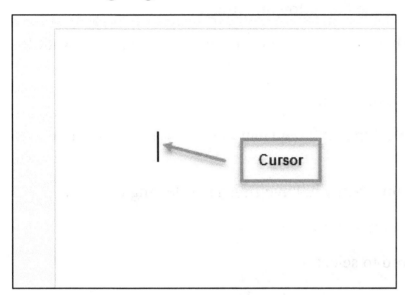

Sample text for students to type: The quick brown fox jumped over the lazy dog.

Selecting Text with the Mouse or Keyboard

To use the keyboard to select text, use the following procedure.

Step 1: Using the arrow keys, place the cursor either at the beginning of the text you want to select, or at the end of the text you want to select.

Step 2: Hold down the shift key while pressing the arrow key to select text in that direction.

The selected text is highlighted.

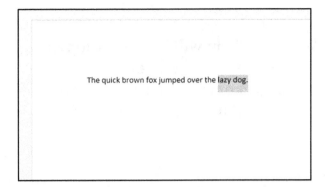

The quick brown fox jumped over the lazy dog.

To use the mouse to select text, use the following procedure.

Step 1: Point the mouse to either the beginning or the end of the text you want to select.

Step 2: Hold the left mouse button down.

Step 3: Move the mouse to select the text. You can move left, right, up and/or down.

Step 4: Let the mouse button up when you have finished selecting the text.

The mouse shortcuts for selecting text are:

- You can double click a word to select it.

- You can click three times on a paragraph to select the whole paragraph.

- You can click to the left of a line to select the whole line.

- You can press Shift while clicking to add to your selection. The selections must be next to each other.

- You can press Control while clicking to add non-congruent text to your selection.

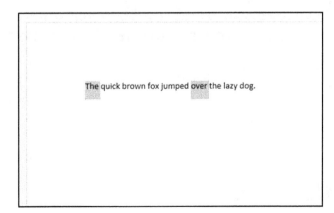

The quick brown fox jumped over the lazy dog.

- Backspace key – Deletes single or multiple characters backwards, or use to delete selected text

- Delete key – Deletes single or multiple characters forwards, or use to delete selected text

- Insert – Place cursor anywhere in text to begin typing. The original text moves to accommodate the inserted text.

- Replace – Select text and begin typing to replace the text.

Dragging and Dropping Text

To drag and drop selected text, use the following procedure.

Step 1: Select the text you want to move.

Step 2: Hold the left mouse pointer down.

Step 3: Move the cursor to the location where you want to move the text. The cursor has an arrow and a small box to indicate that you are moving text.

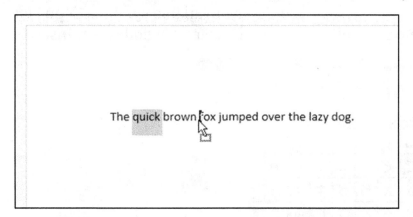

Step 4: Let the mouse button go when the cursor is in the desired location.

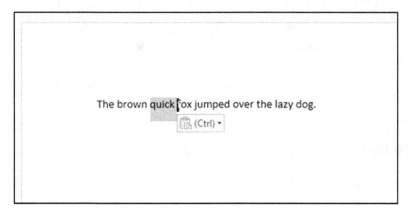

The text remains highlighted in case you want to move it again or continue editing it.

Inserting a Symbol or Number

To insert a symbol, use the following procedure.

Step 1: Select the Insert tab from the Ribbon.

Step 2: Select Symbol.

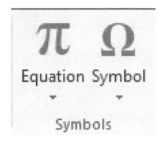

Step 3: Select the symbol from the list, if it is shown. If not, select More Symbols.

Step 4: In the Symbols dialog box, select an option from the Font drop down list and the Subset drop down list to navigate through the available symbols. You can also use the scroll bar on the right. Select the symbol you want and select Insert.

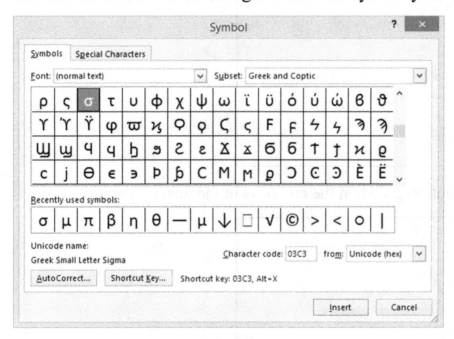

To insert a specially formatted number, use the following procedure.

Step 1: Select the Insert tab from the Ribbon.

Step 2: Select Page Number expanded options.

Step 3: Select Format Page Numbers from the drop-down menu.

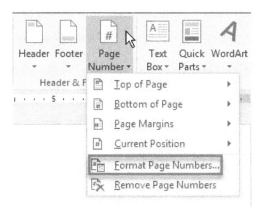

Step 4: In the Page Number format dialog box, enter the Number Format.

Step 5: Select the Page Number Format from the list.

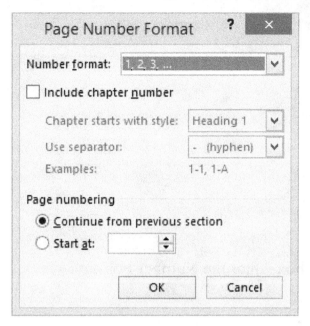

Step 6: Select OK.

Starting a New Page

To insert a page break, use the following procedure.

Step 1: Press Enter to start a new paragraph. This will be important for formatting the document later.

Step 2: Select the Layout Tab on the Ribbon.

Step 3: Select the Breaks tool on the Page Setup Group.

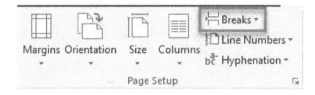

Step 4: Select Page.

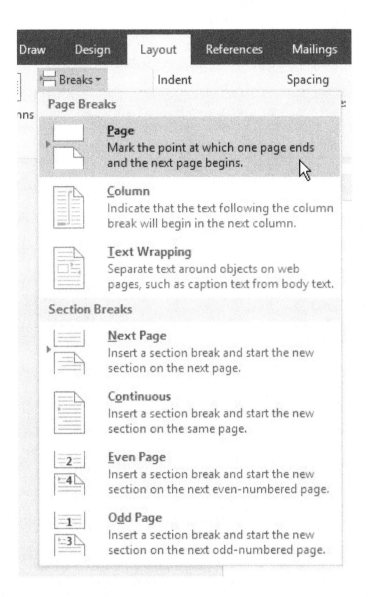

The Word 2016 editing tools make editing your document a breeze. This chapter covers how to cut, copy and paste text, as well as how to undo and redo tasks. It explains how to find and replace text, such as when you want to change a word or phrase throughout your document. It introduces the Word Options dialog box to set default paste options. Finally, it explains how to check your spelling.

Using Cut, Copy, and Paste

To cut and paste text, use the following procedure.

Step 1: Highlight the text you want to cut.

Step 2: Right click the mouse to display the context menu and select cut.

Step 3: Move the cursor to the new location.

Step 4: Right click the mouse to display the context menu and select the Text Only paste option, as illustrated below. Note that the context menu dims so that you can see a preview of your work.

To copy and paste text using the keyboard shortcuts, use the following procedure:

Step 1: Highlight the text you want to cut and press the Control key and the C key at the same time.

Step 2: Move the cursor to the new location.

Step 3: Press the Control key and the V key at the same time.

Using Undo and Redo

To undo their most recent typing or command, use the following procedure.

Step 1: Select the Undo command from the Quick Access Toolbar.

To redo the last command or repeat it, use the following procedure.

Step 1: Select the Redo command from the Quick Access Toolbar.

To find and replace one instance at a time of "Customer Name" in the sample document, use the following procedure.

Step 1: Select Replace from the Editing group on the Home tab of the Ribbon.

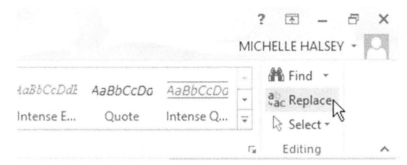

Step 2: In the Find and Replace dialog box, enter the exact text you want to find in the Find what field.

Step 3: Enter the replacement text in the Replace with field.

Step 4: Select Find next to find the next instance of the item.

Step 5: When Word highlights the item, select Replace to delete the "find" item and paste the "replace" item.

Step 6: Select Close when you have finished. Or select Cancel to close the dialog box without making any replacements.

To Replace all instances of an item, use the following procedure.

Step 1: Open the Find and Replace dialog box by selecting Replace from the Ribbon.

Step 2: Enter the exact text you want to find in the Find what field.

Step 3: Enter the replacement text in the Replace with field.

Step 4: Select Replace All.

Step 5: Select Close when you have finished. Or select Cancel to close the dialog box without making any replacements.

Word replaces all instances of the item. If your cursor was not at the beginning of the document, or if you have text selected, Word asks if you want to continue searching at the beginning. When finished, Word displays a message indicating how many replacements were made.

Setting Paste Options

To open the Word Options dialog box for pasting options, use the following procedure.

Step 1: Select the Paste command from the Clipboard group of the Home tab on the Ribbon.

Step 2: Select the Set Default Paste option.

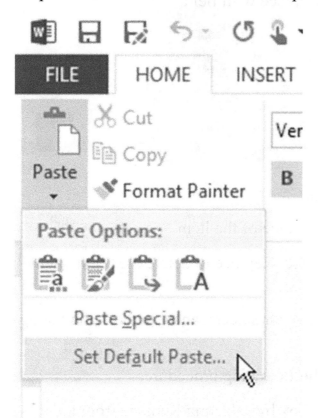

The cut, copy, and paste options on the Word Options dialog box.

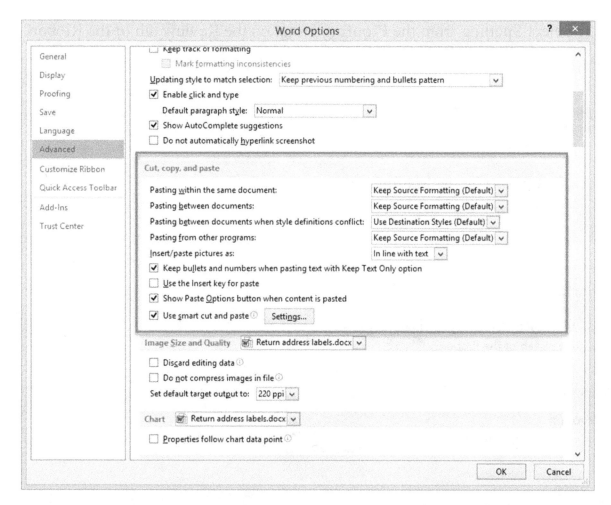

Checking Your Spelling

The following diagram shows the context menu for a misspelled word. The following example uses a misspelling of the word "information."

Step 1: Right click a misspelled word to display the context menu.

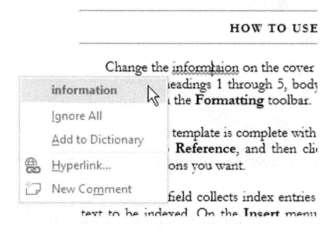

To open the Spelling pane, use the following procedure.

Step 1: Select Spelling from the Proofing group on the Review tab of the Ribbon.

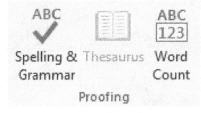

Review the options on the Spelling pane.

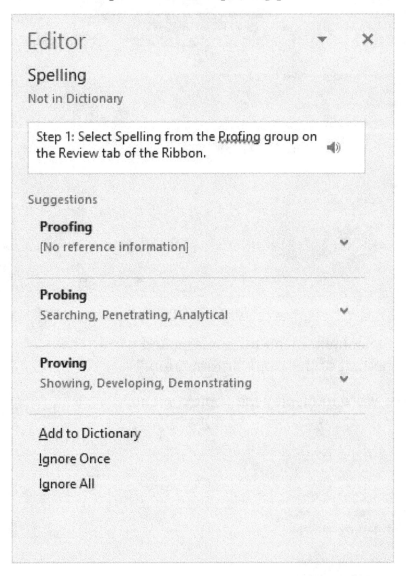

- The Ignore Once button allows you to keep the word as the current spelling, but only for the current location.

- The Ignore All button allows you to ignore the misspelling for the whole document.

- The Add to Dictionary button allows you to add the word to your dictionary for all Word documents.

- The Suggestions area lists possible changes for the misspelling. There may be many choices, just one, or no choices, based on Word's ability to match the error to other possibilities.

- Click the Word once to make a change to the current highlighted word or select the arrow next to the word to see additional options. The Change All button allows you to change the misspelled word to the highlighted choice in the Suggestions area for all instances of the incorrect spelling. You can also add the word to the AutoCorrect options through this menu.

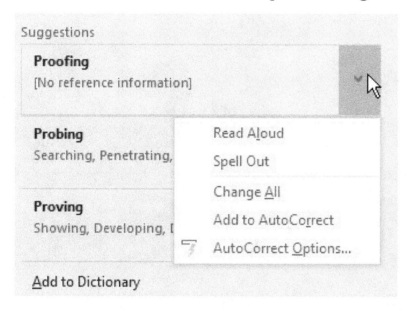

Chapter 5 – Basic Formatting Tasks

Word 2016 allows you to enhance your text in many ways. In this chapter, we will discuss the different types of formatting, as well as cover the most basic types of formatting your words. This includes the font face, size, and color, as well as highlighting and enhancing the text.

Understanding Levels of Formatting

The Font and Paragraph groups on the Home tab of the Ribbon.

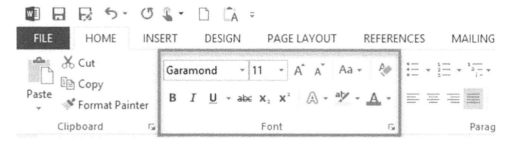

Changing Font Face and Size

To change the font face and size using the Ribbon tools, use the following procedure.

Step 1: Select the text you want to change.

Step 2: Select the arrow next to the current font name to display the list of available fonts.

Step 3: Use the scroll bar or the down arrow to scroll down the list of fonts.

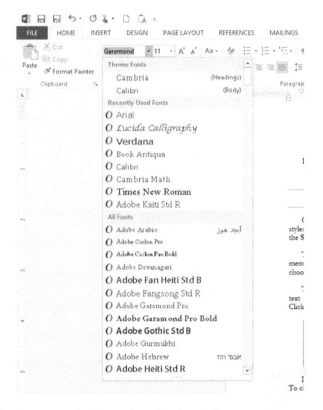

Step 4: Select the desired font to change the font of text.

Step 5: With the text still selected, select the arrow next to the current font size to see a list of common font sizes.

Step 6: Use the scroll bar or the down arrow key to scroll to the size you want and select it. You can also highlight the current font size and type in a new number to indicate the font size you want.

The font context list that appears when you select text, use the following procedure.

Step 1: Select the text you want to change.

Step 2: A very faint context menu appears. Move your mouse over the menu to make sure it stays visible. If you do not see it, you can always right-click the mouse to make it appear.

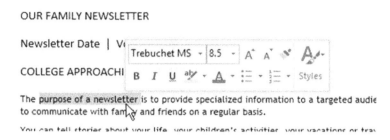

Step 3: Select the new font face or font size just as you would on the Ribbon.

Changing the Font Color

To select a color for their fonts from the gallery, use the following procedure.

Step 1: Select the text you want to change.

Step 2: Select the arrow next to the Font Color tool on the Ribbon to display the gallery. Or select the same tool from the context menu (appears when you select text or by right-clicking).

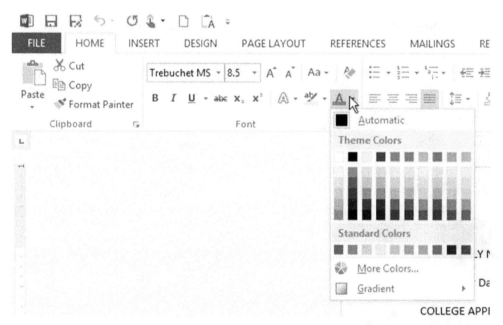

Step 3: Select the color to change the font color.

Use the following procedure to change the Colors of the text.

Step 1: Select the text you want to change.

Step 2: Select the arrow next to the Font Color tool on the Ribbon to display the gallery. Or select the same tool from the context menu (appears when you select text or by right-clicking).

Step 3: Select More Colors to open the Colors dialog box.

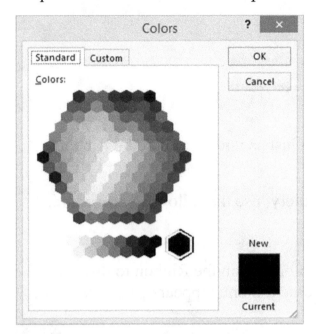

In the Standard Colors dialog box, simply click the color and select OK to use that color.

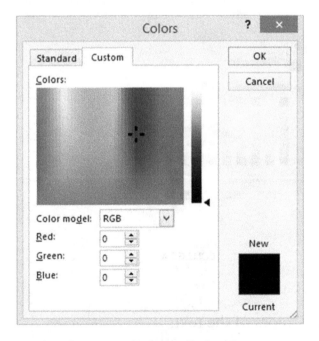

In the Custom Colors dialog box, you can click the color, or you can enter the red, green, and blue values to get a precise color. When you have the color you want, select OK.

Highlighting Text

To highlight text they have already selected, use the following procedure.

Step 1: Select the text you want to highlight.

Step 2: Select the Text Highlight tool from the Ribbon or the formatting context menu. Or select the arrow next to the Text Highlighting tool to choose a highlighting color.

To turn on the highlighting tool to highlight different areas of text, use the following procedure.

Step1: Select the Text Highlight tool from the Ribbon or the formatting context menu. Or select the arrow next to the Text Highlighting tool to choose a highlighting color.

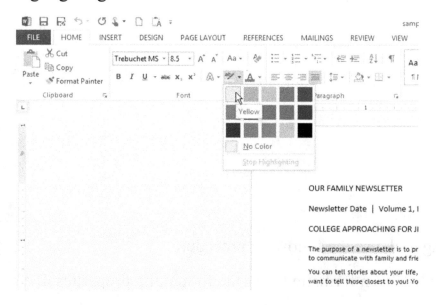

The cursor changes to a highlighting cursor, as illustrated below.

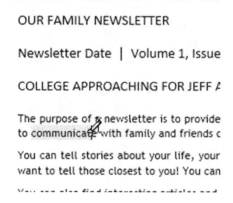

Step 2: Select the text you want to highlight. Word will continue highlighting as many different unconnected pieces of text as you like.

Step 3: To stop highlighting, select the Text Highlight tool again and choose Stop Highlighting. Or just click the Text Highlight tool again.

Adding Font Enhancements

The tools used to add font enhancements.

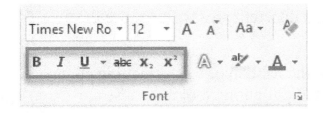

- Bold

- Italic

- Underline

- Strikethrough

- Subscript

- Superscript

Clearing Formatting

To use the clear formatting tool, use the following procedure:

Step 1: Select the text that has been formatted with the formatting properties that you want to remove.

Step 2: Select the Clear Formatting tool.

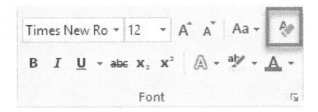

Chapter 6 –Formatting Paragraphs

Paragraph formatting controls the look and feel of an entire paragraph. In this chapter, we will discuss how to change the spacing of your text, both the line spacing and the space in between paragraphs. We will also address setting the alignment and using tabs and indents. We will also practice using bullets and numbering the document and learn how to add borders and shading to the text. Finally, we will look at the Paragraph dialog, where you can format many aspects of your paragraph at once.

Changing Spacing

To adjust the line spacing using the Line Spacing tool on the Ribbon, use the following procedure.

Step 1: With your cursor, anywhere in the paragraph you want to adjust (the text does not have to be selected), select the Line and Paragraph spacing tool from the Ribbon.

Step 2: Select one of the following options:

- 1.0 – single spacing

- 1.15 – provides a little more space than single spacing

- 1.50 – One and a half line spacing

- 2.0 – double spacing

- 2.5 – two and a half line spacing

- 3.0 – triple spacing

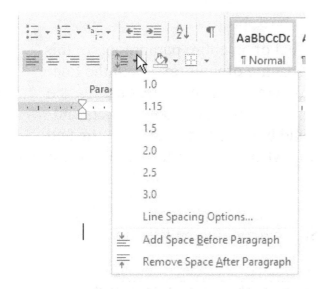

To add or remove space before or after a paragraph, use the following procedure.

Step 3: With your cursor, anywhere in the paragraph you want to adjust (the text does not have to be selected), select the Line and Paragraph spacing tool from the Ribbon.

Step 4: The Paragraph spacing options listed are based on your current settings. You can choose one of the following to add or remove space before or after your paragraph:

- Add Space Before Paragraph

- Remove Space Before Paragraph

- Add Space After Paragraph

- Remove Space After Paragraph

The amount added by default is usually 12 points. To add more, you will need to use the Paragraph dialog box.

Setting the Alignment

To adjust the alignment for the paragraph, use the following procedure.

Step 1: With your cursor, anywhere in the paragraph you want to adjust (the text does not have to be selected), select the desired alignment tool from the Ribbon. You can also select multiple paragraphs by selecting the text.

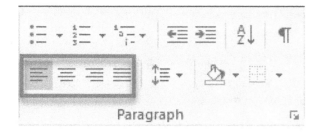

Using Indents and Tabs

To add a whole paragraph indent, use the following procedure.

Step 1: With your cursor, anywhere in the paragraph you want to adjust (the text does not have to be selected), select the Indent tool from the Ribbon. You can also select multiple paragraphs by selecting the text.

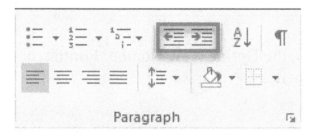

Adding Bullets and Numbering

To create a simple bulleted or numbered list, use the following procedure.

Step 1: Select the paragraphs you want to turn into a bulleted or numbered list.

Step 2: Select the Bullets or Numbering tool from the Ribbon.

The Bullet Library

Step 1: Select the arrow next to the Bullets tool or the Numbering tool on the Ribbon to view the library options.

Step 2: Select an option to create a list with that option.

To open the Define New Bullet dialog box, use the following procedure.

Step 1: Select the arrow next to the Bullets tool on the Home Ribbon.

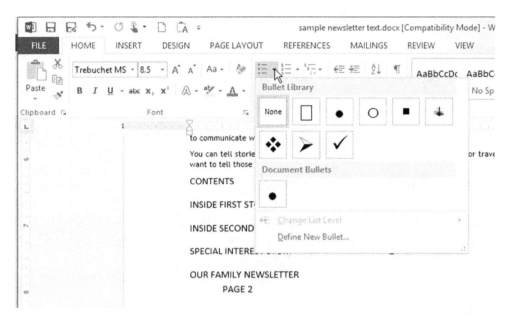

Step 2: Select the Define New Bullet option, the Define New Number Format, or the Set Numbering Value option from the menu.

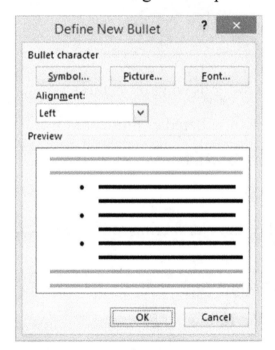

Step 3: Select the bullet options. You can choose a symbol, picture, or font and then select the alignment.

Step 4: Click Ok.

The Numbering Library

Step 1: Select the arrow next to the Numbering tool on the Home Ribbon to view the library options.

Step 2: Select an option to create a list with that option.

To open the Define New Numbering Format dialog box, use the following procedure:

To open the Set Numbering Value dialog box, use the following procedure.

Step 1: Select the arrow next to the Numbering tool on the Home Ribbon.

Step 2: Select Define New Number Format.

Step 3: Make the desired formatting changes.

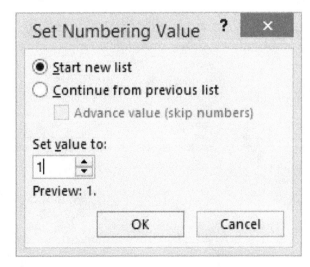

Step 4: Click Ok.

To open the Set Numbering Value dialog box, use the following procedure.

Step 1: Select the arrow next to the Numbering tool on the Home Ribbon.

Step 2: Select Set Numbering Value.

Step 3: Make the desired formatting changes.

Step 4: Click Ok.

The Shading and Border tools on the Ribbon are illustrated below.

To add shading to selected paragraphs, use the following procedure.

Step 1: Select the paragraphs you want to shade. If you only want to shade one paragraph, your cursor can be anywhere in the paragraph without selecting it.

Step 2: Select the color from the Shading tool on the Ribbon. The Shading tool includes the same gallery of colors as previously introduced.

To add borders to selected paragraphs, use the following procedure.

Step 1: Select the paragraphs you want to border. If you only want to put borders on one paragraph, your cursor can be anywhere in the paragraph without selecting it.

Step 2: Select the border you want to use from the Borders tool on the Ribbon. The Borders tool includes several options for borders. Some of the options only apply for tables.

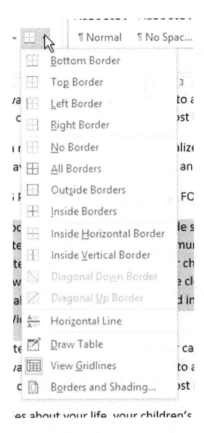

The Borders and Shading dialog box.

Step 1: Open the Borders and Shading dialog box by selecting Borders and Shading from the Borders tool on the Ribbon.

The Borders tab of the Borders and Shading dialog box is illustrated below. The Borders tab allows you to format a board around a range of text or an image on a page.

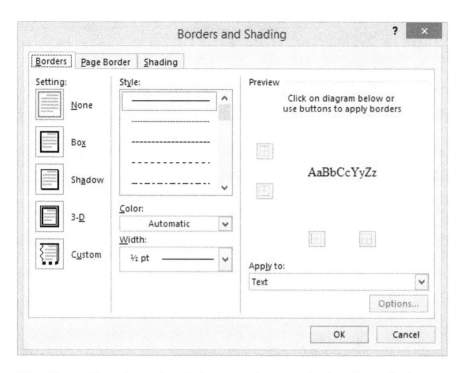

The Page Border tab of the Borders and Shading dialog box is illustrated below. Page Borders allows you to place a boarder around an entire page or the entire document.

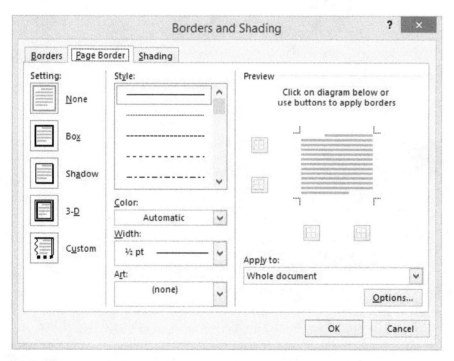

The Shading tab of the Borders and Shading dialog box is illustrated below. The Shading tab allows you to set shading elements around a block of text.

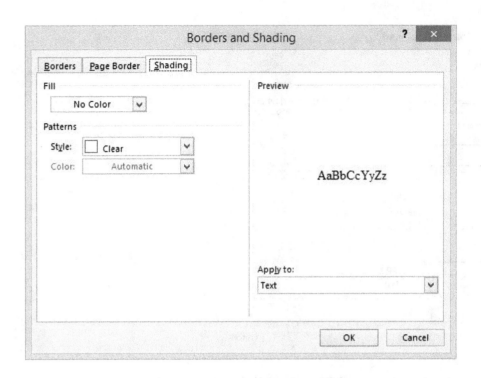

Chapter 7 – Advanced Formatting Tasks

This chapter introduces some of the more advanced formatting tasks for formatting your text in Word 2016. This chapter covers changing the case (capitalization) of words. You will also learn to use the format painter, to quickly format words to match others. This chapter introduces the Font dialog box for formatting several properties of your font at once. Finally, you will learn how to clear your formatting choices if you change your mind about the formatting.

Changing Case

To change the case, use the following procedure.

Step 1: Select the text you want to change.

Step 2: Select the Case tool from the Font group of the Home tab on the Ribbon.

Step 3: Select the Case option from the drop-down list.

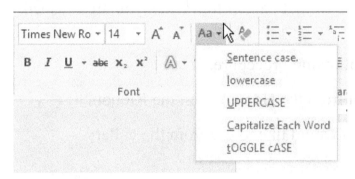

Using the Format Painter

To use the Format Painter, use the following procedure.

Step 1: Select the text that has been formatted with the formatting properties that you want to copy.

Step 2: Select the Format Painter tool.

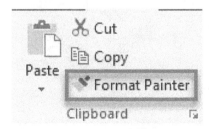

The cursor changes to a Format Painter cursor, as illustrated below.

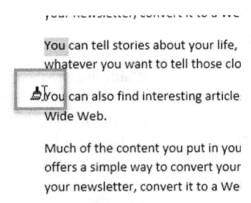

You can tell stories about your life, whatever you want to tell those clo

You can also find interesting article Wide Web.

Much of the content you put in you offers a simple way to convert your your newsletter, convert it to a We

Step1: Select the text you want to format with the same properties.

The cursor returns to normal after applying the formatting properties once. You can always repeat the process to format more text with the same properties.

If you double-click the format painter tool before applying it to text, you can use it several times in a row. Just click the format painter tool when you are finished.

Creating Multilevel Lists

To create a multilevel list, use the following procedure.

Step 1: Select the Multilevel list tool from the Home tab on the Ribbon.

Step 2: Select the type of list that you would like to use from the gallery.

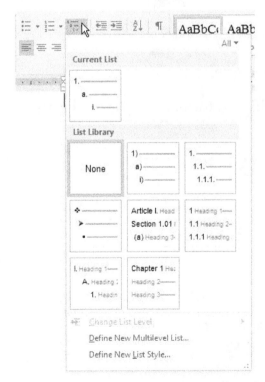

Step 3: Begin typing the list. In this example, you can use simple text, like "level 1" and "level 2".

Step 4: Press Enter to move to the next item.

Step 5: If the item is not automatically formatted/numbered properly, select Change List Level from the Multilevel list drop down list.

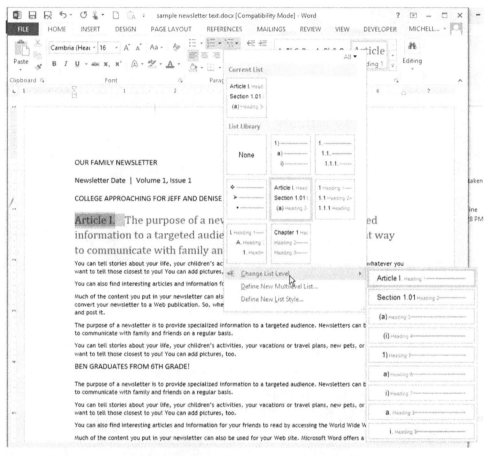

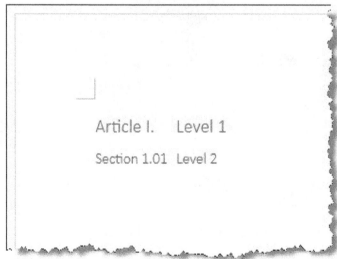

To open the Font dialog box, use the following procedure.

Step 1: Select the text you want to format.

Step 2: Select the square at the bottom right corner of the Font group in the Ribbon.

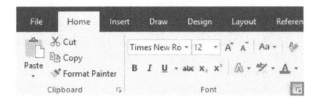

The following diagrams show the Font dialog box.

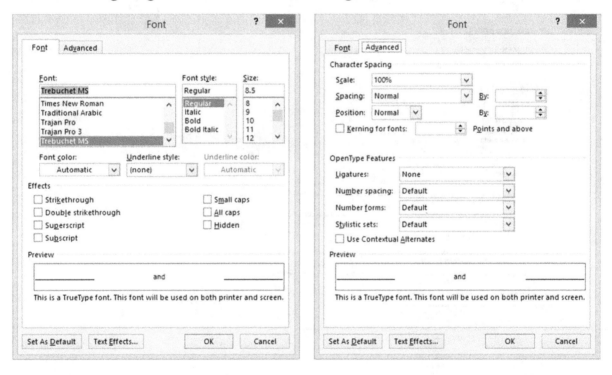

The following diagram shows the Set as Default dialog box.

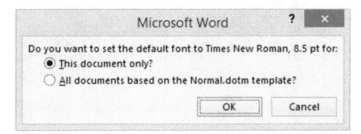

Review the Text Effects dialog box options. Click the icons at the top or the arrows to expand the available options for each item.

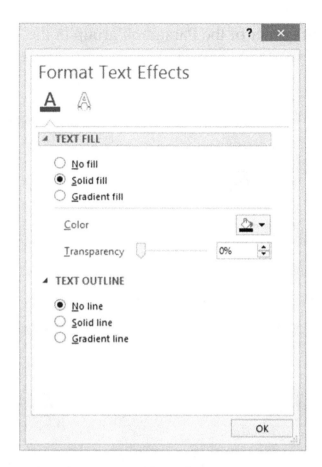

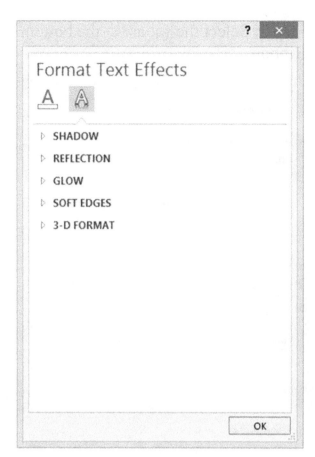

Using the Paragraph Dialog

Use the following procedure to use the Paragraph dialog box:

Step 1: With your cursor, anywhere in the paragraph you want to adjust (the text does not have to be selected), select the Line and Paragraph spacing tool from the Ribbon.

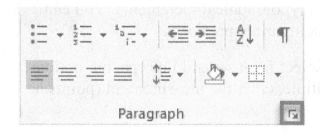

Step 2: Select the square at the bottom right corner of the Paragraph group in the Ribbon.

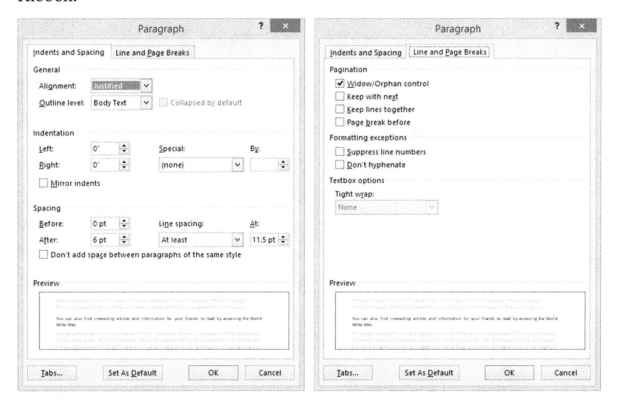

- The Special field allows you to select a first line only or hanging indent. Enter the measurement for the special indent in the By field. Check the Mirror indents to have the indent on both the left margin and the right margin by the same amounts.

- You can use the up and down arrows to adjust the indentation and spacing options. The arrows adjust the points in typographical increments. You can also enter any number to adjust the spacing more precisely.

- The Line Spacing field allows you to select from several line spacing options. If you select At Least, Exactly, or Multiple, enter the measurement (points or lines) in the At field.

- You can preview your selections at the bottom of the dialog box.

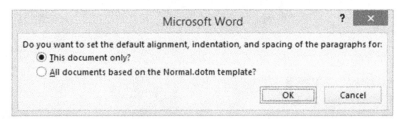

- Click the Tabs button to format the tab stops and positioning

Chapter 8 – Working with Styles

Styles are a powerful formatting tool to take your Word 2016 document to the next level. Styles help provide consistency. They are also useful if you want to use certain advanced features like generated tables of contents. This chapter introduces styles and themes to help make your documents look great.

About Styles

The implications of using styles may not be apparent in shorter documents, but they are a great time saver for longer documents. They also help ensure that your document is consistently formatted. Styles also provide an easy way to easily change the look of the whole document if styles have been applied appropriately.

Once you have applied Heading styles to your document, the Navigation pane will also help you to quickly access different parts of the document based on the heading styles. Styles are a great time saver!

Applying a Style

Use the Style gallery to apply a paragraph or character style.

Step 1: Select the text you want to format, or simply place your cursor in the word or paragraph you want to format.

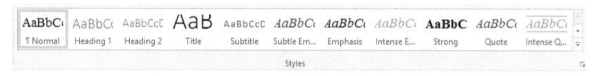

Step 2: Open the Style Gallery by clicking the down arrow next to the styles shown in the Styles group.

Step 3: Select the desired style to apply it to the current word or paragraph.

The Apply Styles dialog box.

Step 1: Select the text you want to format, or simply place your cursor in the word or paragraph you want to format.

Step 2: Open the Apply Styles dialog box by clicking the down arrow next to the styles shown in the Styles group, and selecting Apply Styles from the menu.

Step 3: To apply a style using the Apply Styles dialog box, simply begin typing the name of the style and press Enter when the desired style is displayed. Or use the drop-down list to select the style.

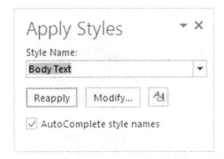

Changing the Theme

To change the theme, use the following procedure.

Step 1: Select the Design tab on the Ribbon.

Step 2: Select the Themes tool from the Ribbon to see the options.

Step 3: Select a Theme from the list.

Changing Theme Colors and Fonts

To change the theme colors or fonts, use the following procedure.

Step 1: Select the Design tab on the Ribbon.

Step 2: Select the Colors tool or the Fonts tool from the Ribbon to see the options.

Step 3: Select an option from the menu to change the color set or font set for the document.

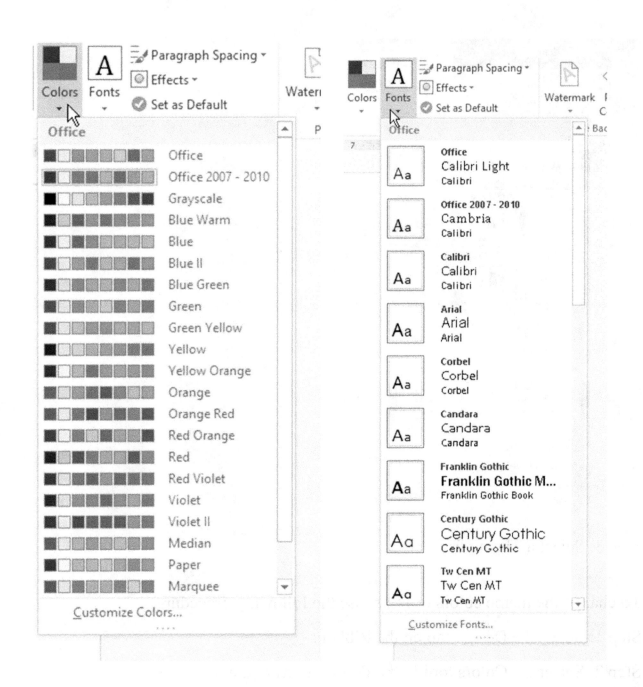

Chapter 9 – Formatting the Page

You have your text and paragraphs looking great, but what about the page? This chapter covers the basics of page formatting. You will learn how to format text into columns, how to change the orientation from portrait to landscape, how to add a page color or border, and how to add headers and footers.

Formatting Text as Columns

To create columns, use the following procedure.

Step 1: Select the Layout tab from the Ribbon.

Step 2: Select the Columns tool.

Step 3: Select the number or layout of columns that you want to use.

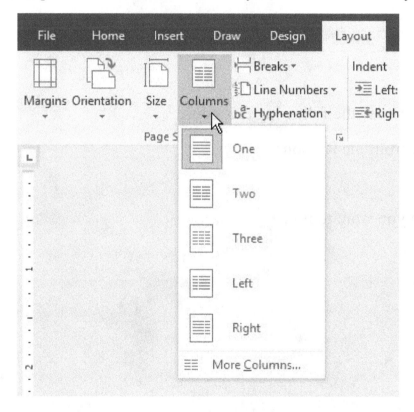

Click More Columns to open the Columns dialog box.

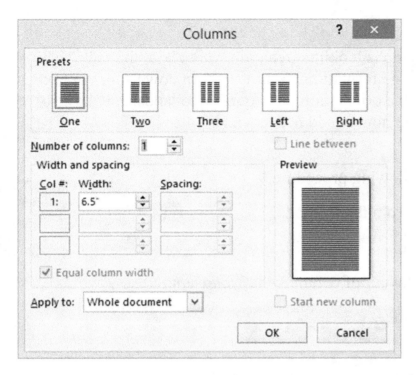

Changing Page Orientation

To change the page orientation, use the following procedure.

Step 1: Select the Layout tab from the Ribbon.

Step 2: Select Orientation.

Step 3: Select the orientation you want to use.

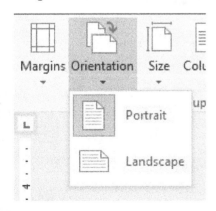

Changing the Page Color

To add color to the page, use the following procedure.

Step 1: Select the Design tab from the Ribbon.

Step 2: Select the Page Colors tool.

Step 3: Select a color from the gallery.

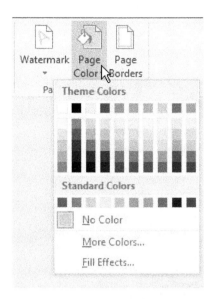

Click Fill Effects to access the Fill Effects dialog boxes.

The Gradient tab of the Fill Effects options allow you to apply gradients to a shape or document component.

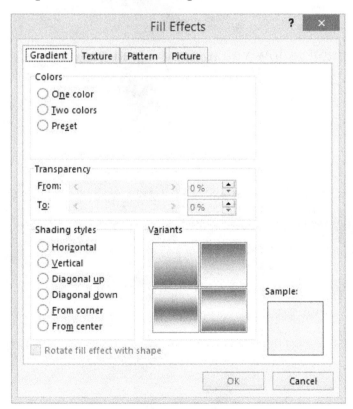

The Texture tab of the Fill Effects options allow you to apply textures to a shape or document component.

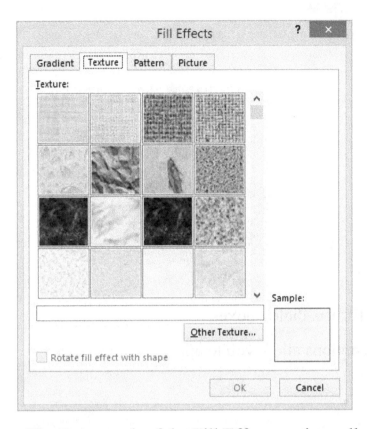

The Pattern tab of the Fill Effects options allow you to apply a pattern to a shape or document component.

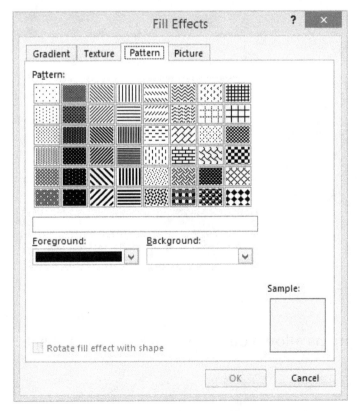

The Picture Tab of the Fill Effects option allows you to apply a picture to a shape or other document compnent.

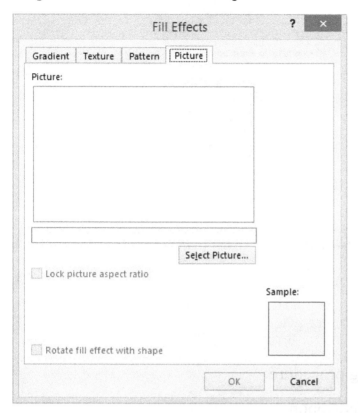

Adding a Page Border

To add a page border, use the following procedure.

Step 1: Select the Design tab from the Ribbon.

Step 2: Select Page Borders.

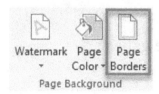

Step 3: Select the type of border using the Setting or Style options. You can select a color, the width, and even art. Click the diagram to create a custom border.

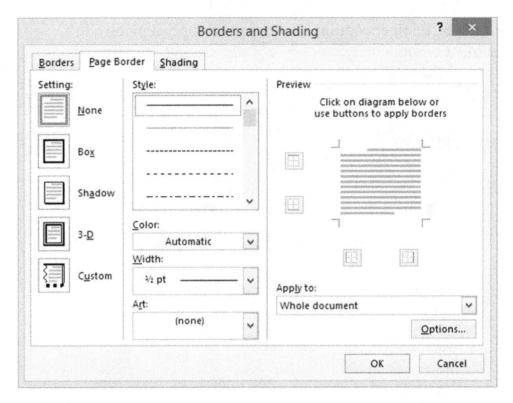

Adding Headers and Footers

To insert a header, use the following procedure.

Step 1: Select the Insert tab from the Ribbon.

Step 2: Select Header or Footer.

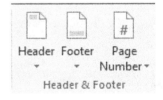

Step 3: Select the type of header that you want to use from the gallery.

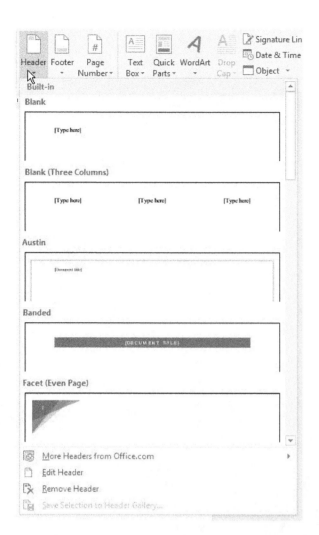

The following diagram shows the Header & Footer Tools Design tab.

Chapter 10 – Sharing Your Document

Now that your document is ready, it is time to share it! First, you will learn how to preview and print your document. When you have saved your document to the cloud, you can invite people, which sends a link so that you can share the document. You can also get a link to share the people, which send groups of people (such as when you do not know everyone's email address). Finally, you will learn how to email the document.

Previewing and Printing Your Document

To open the Print tab of the Backstage View to preview the document, use the following procedure.

Step 1: Select the File tab on the Ribbon.

Step 2: Select the Print tab in the Backstage View.

Note the buttons on the Print tab of the Backstage View.

- The Print button allows you to print the document using the current settings.

- The Copies field allows you to print one or more copies of the document.

- The Printer allows you to select a different printer. The printer properties link allows you to set the properties for that printer.

- The Settings tool allows you to select different pages of your document. You can even print document properties, such as a list of styles used in the document.

- The pages field allows you to specify a custom page range to print.

- The other settings control additional settings for print, such as one or two-sided printing, whether multiple copies are collated, the orientation, the paper size, the default page margins, and how many pages to print per page.

- There is also a link to the Page Setup dialog box.

Sharing Your Document

To invite people to the document, use the following procedure.

Step 1: Select the File tab from the Ribbon to open the Backstage view.

Step 2: Select the Share tab.

Step 3: Select Share with People.

Step 4: Click the Save to Cloud button. Note, you may need to log in to your account to share your document via the cloud.

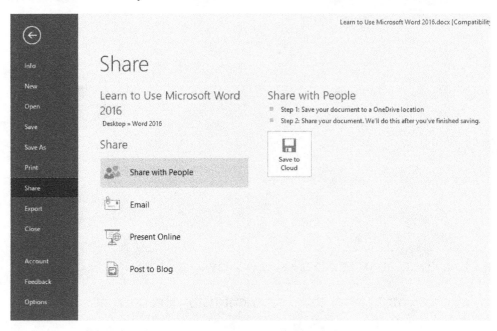

Step 5: Save the document to the cloud drive and then share the file with other recipients.

Step 6: Enter the names or email addresses for the people that you want to invite.

Step 7: Enter a message to include with the invitation.

Step 8: If desired, check the require user to sign-in before accessing document box to enhance the security of your document.

Step 9: Select Share.

To get a link for the document, use the following procedure.

Step 1: Select the File tab from the Ribbon to open the Backstage view.

Step 2: Select the Share tab.

Step 3: Select the Email option.

Step 4: Select Send a Link.

Step 4: Select the Create Link button next to View Link or Edit Link (or both), depending on what type of editing rights you want to provide. You can copy the link and paste it to another location, such as an email or a blog page.

Step 5: If you want to remove the sharing rights, select Disable Link.

E-Mailing Your Document

To email an attachment or send a link, use the following procedure.

Step 1: Select the File tab on the Ribbon.

Step 2: Select the Share tab in the Backstage View.

Step 3: Select Email.

Step 4: Select Send as Attachment or Send a Link.

Step 5: Outlook opens with an email started.

- If you select Send as Attachment, the name of the document is used as the subject and the document is already attached to the email. Enter the email addresses and any personal message you want to include.

- If you select Send a Link, the name of the document is used as the subject and the link is included in the body message of the email. Enter the email addresses and any personal message that you want to include.

Customizing the Word window is a powerful way to become more productive when working with documents. The first concept we will cover in this chapter is the Zoom feature, which allows you to focus in on details, or else zoom out to see the entire document. We will present an overview of the various document views available in Word. This chapter will explain how to arrange multiple windows to see more than one document at a time. We will also cover how to split a document, so that you can see more than one location in a long document at one time. You will learn some advanced uses of the Navigation pane. Finally, you will learn how to customize both the Ribbon and the Quick Access toolbar, so that your most used commands are always at your fingertips.

Using Zoom

Use the following procedure to zoom using the Status bar.

Step 1: Click the Zoom slider options to use Zoom.

- Click + to Zoom in (up to 500% of the normal view).

- Click – to Zoom out (down to 10% or small enough to see many pages at once).

- Drag the slider toward the + to zoom in or toward the – to zoom in.

- Click the current percentage to open the Zoom dialog box.

To modify and review the options of the Zoom dialog box, use the following procedure.

Step 1: Select the View Ribbon.

Step 2: Select the Zoom button in the Zoom menu to open the Zoom dialog box.

Step 3: Select a Zoom to option.

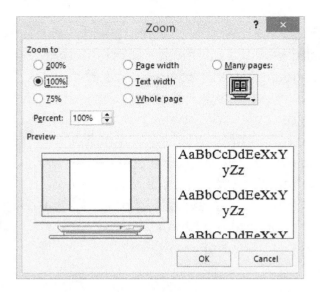

- 200% is twice the print size.

- 100% is the print size.

- 75% is smaller than the print size.

- Page Width scales the view to the width of the page.

- Text Width scales the view to the width of the text.

- Whole page scales the view to show the whole page in one screen.

- Percent allows you to enter a precise zoom percentage. Enter a percentage or use the up and down arrows.

- Many Pages allows you to select how many pages to view in the screen. Click the arrow to select the number of pages to include.

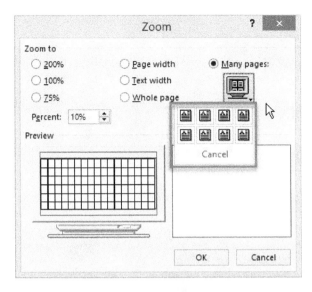

Step 4: Select OK to apply the zoom. Or select Cancel to close the Zoom dialog box without changing the zoom.

An Overview of Word's Views

The Read Mode opens a view for reading a document on the screen. There is a minimized Ribbon with access to the Backstage View, find and search tools, and additional tools on the View tab for reading your document or switching back to editing view. To switch to Read Mode, select the View tab from the Ribbon and then select Read Mode.

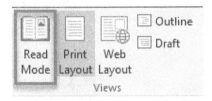

Or select the Read Mode icon from the Status Bar.

Tools available when working in Read Mode.

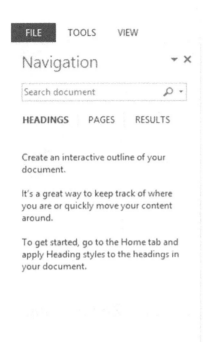

The Print Layout view makes the screen look exactly like the document will look when it prints. To switch to Print Layout view, select the View tab from the Ribbon. Select Print Layout. Or select the Print Layout icon from the Status Bar.

The third view available in Word 2016 is the Web Layout view. This view simulates what your document would look like on the Internet. To switch to Web Layout view, select the View tab from the Ribbon. Select Web Layout. Or select the Web Layout icon from the Status Bar.

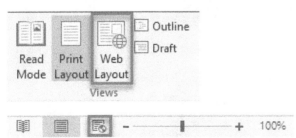

The fourth view available in Word 2016 is the Outline view. Outline view is a special view for working with levels (or paragraphs that have a Heading level style applied). This view can help you get a handle on the structure and organization of your document.

Click the View tab and click Outline to switch to Outline view.

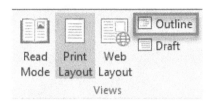

While in Outline mode, the Outlining Ribbon will open which gives you options for outlining the document.

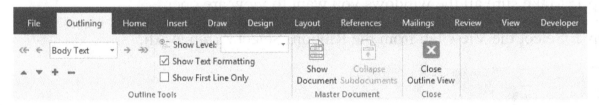

The fifth view available in Word 2016 is the Draft view. Draft view removes all extra white space (such as margins) and pictures. Draft view allows you to scroll much more quickly through a long document. It also allows you to see more on the screen at one time without having to adjust the zoom too small. Select the View tab and then click Draft to switch to Draft view.

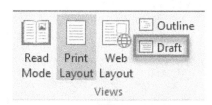

Arranging Windows

To arrange open windows, use the following procedure.

Step 1: Make sure all the windows you want to view are open.

Step 2: Select the View tab from the Ribbon. Select Arrange All.

Word resizes the windows to each take a percentage of the screen; depending on how many documents you have open. The original document is on top.

Select the Maximize icon from the top right corner of one of the windows to return the window to its previous size and position.

Splitting a Document
To split the view, use the following procedure.

Step 1: Select the View tab from the Ribbon. Select Split.

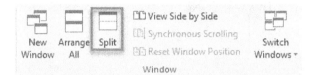

Step 2: Word splits the view into two windows. You can drag the resize line to make one window smaller or larger.

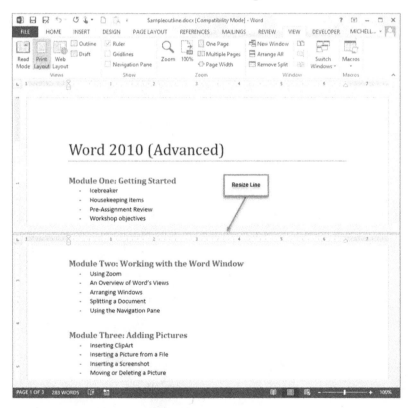

Step 3: The two windows include separate rulers and scroll bars, but not a separate Ribbon. Place your cursor in the appropriate window to apply a command to that section of the document. All the commands will work in either window.

Step 4: To remove the split, select the View window from the Ribbon. Select Remove Split.

To open the Navigation pane, use the following procedure.

Step 1: Select the View tab from the Ribbon. Check the Navigation Pane box.

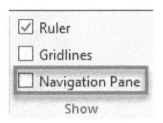

To navigate to another section using the Navigation pane, use the following procedure.

Step 2: Click a heading to go to that section of the document.

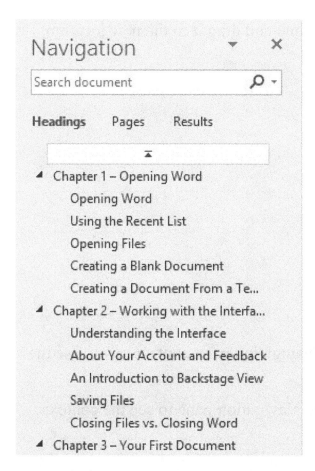

To rearrange the sections in a document using the Navigation pane, use the following procedure.

Step 3: Click a heading in the Navigation pane and drag it to the new location.

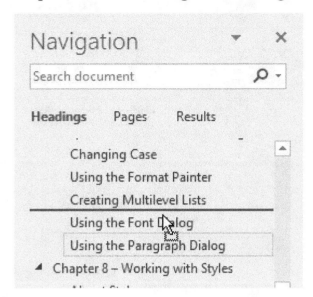

To review the options and settings in the Navigation pane context menu, use the following procedure.

Step 1: Right click a section heading in the Navigation pane to see the context menu.

Customizing the Ribbon and the Quick Access Toolbar

To customize the Ribbon, use the following procedure.

Step 1: Select the File tab from the Ribbon to open the Backstage View.

Step 2: Select Options.

Step 3: Select Customize Ribbon from the left side.

In the left column, under Choose Commands From, Word lists the commands available in the application. You can choose a different option from the Choose Commands From drop down list to change which options are shown or how they are sorted.

The right column shows the available tabs on the Ribbon.

Step 1: To customize the Ribbon, select the command that you want to change on the left column. Select Add. You may need to create a Custom Group before you can add a command.

- Select the Tab where you want the group to appear.

- Select New Group.

- Enter the Group name.

You can also remove commands or rearrange them on the right column.

When you have finished, select OK.

The procedure is similar when adding a command to the Quick Access Toolbar, except that you do not need to add a custom group for commands.

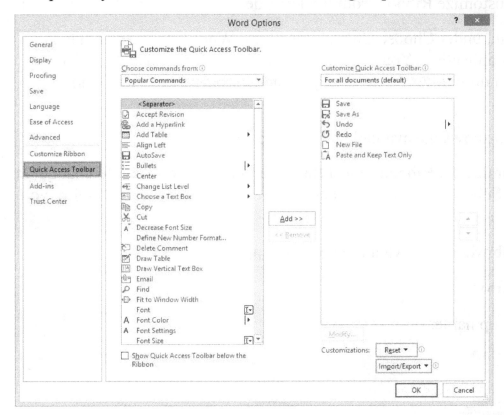

Chapter 12 – Advanced Editing and Formatting Tasks

This chapter will help you understand some more sophisticated tools to format your text, such as the character borders and shading, enclosing characters, and text effects and typography options. You will also learn how to use the phonetic guide to help you readers with pronunciation. First, we will start off with introducing the Office Clipboard to help you with multiple copy and paste tasks.

Using the Office Clipboard

To open the Clipboard Task pane, use the following procedure.

Step 1: The Home tab of the Ribbon, select the icon next to Clipboard.

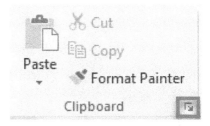

The Clipboard pane opens, displaying any items you have cut or copied in this Word 2016 session (or the 24 most recent). A sample is illustrated below.

To paste using the Office Clipboard Task pane, use the following procedure.

Step 1: Place the cursor where you want to paste text from the clipboard.

Step 2: Click the item in the Clipboard task pane that you want to paste.

Using the Phonetic Guide

If you do not see the Phonetic Guide icon, you may need to enable an Asian language in Word. Go to the File ribbon, click Options, and select Language.

Select an Asian Language from the Language drop-down and click Add. You may need to restart Word.

To use the Phonetic Guide, use the following procedure.

Step 1: Highlight the word that you want to enhance using the Phonetic Guide.

Step 2: Select the Phonetic Guide icon from the Font group on the Home tab of the Ribbon.

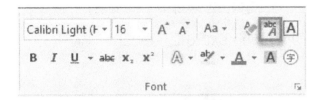

Step 3: Enter (or paste) the Ruby text.

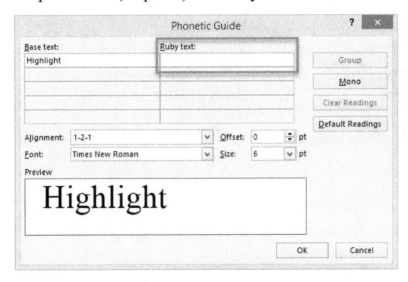

Step 4: There are additional options to change the ruby text alignment, offset, font, and size. You can group the word or illustrate the pronunciation using Mono to separate the letters.

Step 5: Select OK when you have finished. The word is highlighted in the document with the ruby text above.

Using Character Borders and Shading

Use the following procedure to apply borders or shading to text.

Step 1: Select the text that you want to enhance.

Step 2: Select the Character Border tool or the Character Shading Tool.

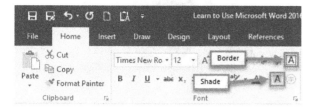

The following example has both borders and shading applied to the selected text.

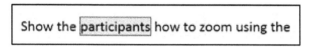

Enclosing Characters

To enclose characters, use the following procedure.

Step 1: Select the character that you want to enhance.

Step 2: Select the Enclose Characters command from the Font group on the Home tab of the Ribbon.

Step 3: In the Enclose Characters dialog box, select the Style of symbol you want to use.

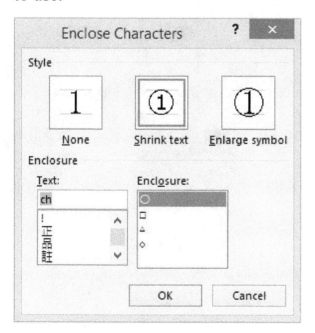

Step 4: The character you selected is the default option under Text. You can select another item from the list, if desired.

Step 5: Select the type of Enclose from the list.

Step 6: Select OK.

Using Text Effects

To apply text effects or typography options, use the following procedure.

Step 1: Select the text that you want to enhance.

Step 2: Select the Text effects command from the Font group on the Home tab of the Ribbon.

Step 3: Select the option that you want to use.

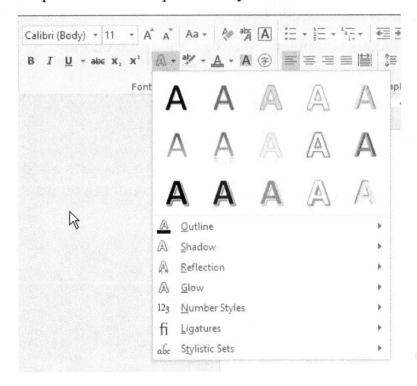

Chapter 13 – Working with Illustrations

This chapter will help you learn how to insert and work with pictures and other illustrations in your document. We will cover pictures from files as well as online pictures, WordArt, shapes, and screenshots. You will also learn how to move and delete the illustrations.

Inserting a Picture from a File

To insert a picture from a file, use the following procedure.

Step 1: Select the Insert tab from the Ribbon.

Step 2: Select Picture.

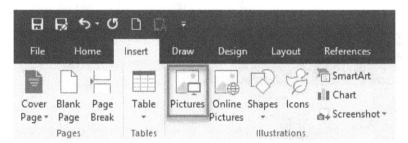

Step 3: Navigate to the location of the file and highlight the file you want to insert.

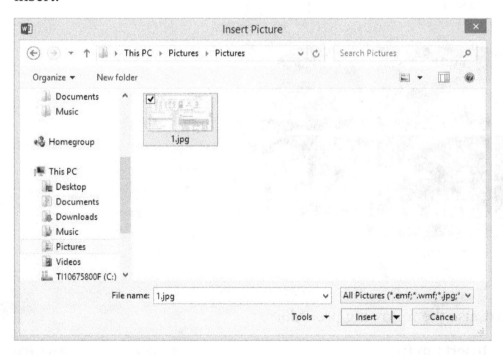

Step 4: Select Insert.

Word inserts the picture.

To insert a clip art, use the following procedure.

Step 1: Select the Insert tab from the Ribbon.

Step 2: Select Online Pictures.

Step 3: In the Insert Pictures dialog box, select the place where you want to search for images.

Step 4: Enter a search term. Press Enter to begin searching.

Step 5: Word displays the matching images. To insert one, double-click it or highlight it and select Insert.

Adding WordArt

To insert WordArt, use the following procedure.

Step 1: Select the Insert tab from the Ribbon.

Step 2: Select WordArt.

Step 3: Select the style you would like to use.

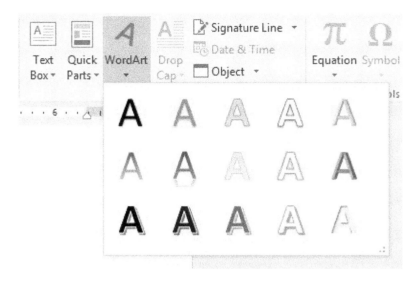

Step 4: Word inserts the text box with the placeholder text highlighted. Enter your own text to replace the placeholder text.

Notice the icon to the right of the text box. Click it to see your layout options.

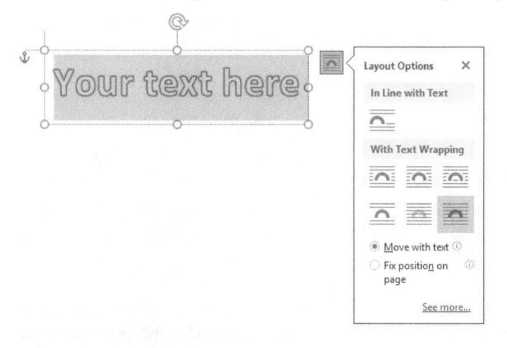

To insert a built-in shape, use the following procedure.

Step 1: Select the Insert tab from the Ribbon.

Step 2: Select Shapes.

Step 3: Select the shape that you want to use.

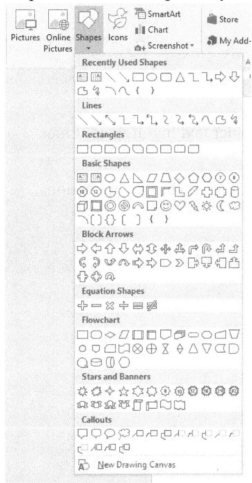

Step 4: Hold down the mouse button from the location in the document where you want to place the top left of the shape. Drag down and to the right it until the shape is the desired shape and size. The cursor changes to a cross while you are drawing.

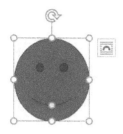

To draw with one of the freehand shapes, use the following procedure

Step 1: Select the Insert tab from the Ribbon.

Step 2: Select Shapes.

Step 3: The freehand drawing tools are in the Lines section. Select either the closed freehand shape (Freeform) or the open freehand shape (Scribble).

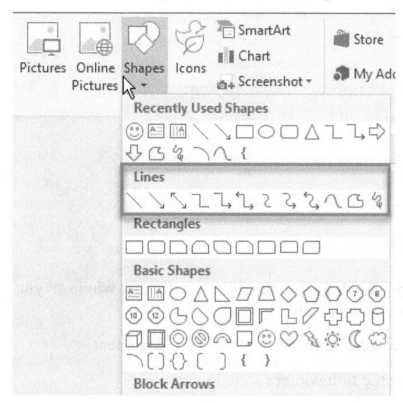

Step 4: Begin drawing. If you are using the Freeform tool, Word will close the shape when you click close to your starting point. If you are using the Scribble tool, Word will finish the shape when you stop dragging the mouse.

Inserting a Screenshot

To insert a full-size screenshot, use the following procedure.

Step 1: Select the Insert tab from the Ribbon.

Step 2: Select Screenshot.

Step 3: The Screenshot gallery includes a thumbnail image of other windows you have open. Select the image that you want to insert.

Word inserts the image and may scale it to the width of your document.

To insert a screen clipping, use the following procedure.

Step 1: Make sure that the area of the screen you want in your document is ready to capture. Word will automatically return to the previous window for a screen clipping.

Step 2: Select the Insert tab from the Ribbon.

Step 3: Select Screenshot.

Step 4: Select Screen Clipping.

Step 5: Drag the mouse to capture the area of the screen that you want to insert in your presentation. The screen is slightly greyed out, except for the area you are capturing.

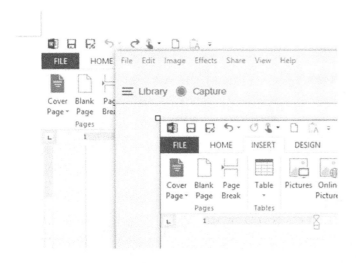

Step 6: When you release your mouse, Word inserts the screen clipping into the document at the current cursor position.

Moving or Deleting a Picture

To move a picture, use the following procedure.

Step 1: Select the picture you want to move.

Step 2: Drag the mouse until the picture is in the desired location. Word displays a small rectangle by the cursor to show an object is being moved. There is a small line showing where the picture will be moved.

Release the mouse to drop the picture in the new location.

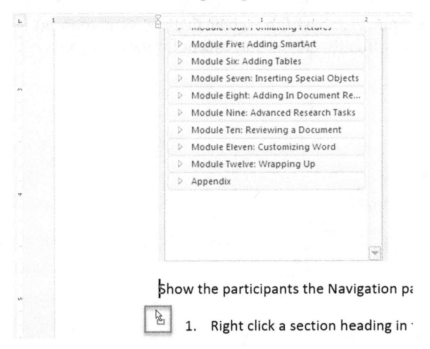

To delete a picture, use the following procedure.

Step 1: Select the picture you want to delete.

Step 2: Press the Delete key on the keyboard.

Chapter 14 – Formatting Pictures

In this chapter, you will learn how to use the Picture Tools Tab. Removing a picture's background is a great technique to add professionalism to your documents. You will also learn how to add a border to a picture. This chapter also explains how to add artistic effects and change a picture's position and text wrapping. You will also learn how to use the selection pane for selecting an object when multiple objects are layered.

Using the Picture Tools Tab

To use the Picture Tools tab, use the following procedure.

Step 1: With a picture selected on the document, select the Picture Tools/Format tab from the Ribbon. You can use the other tabs while working with a picture, and this tab will still be available.

Adding a Border

To add a border to a picture, use the following procedure.

Step 1: Select the picture to which you want to add a border.

Step 2: Select the Picture Tools/Format tab from the Ribbon.

Step 3: Select Picture Border.

Step 4: Select a color from the gallery to use or select More Outline Colors to choose a Standard or Custom color as you've seen in other Word color galleries.

Step 5: Select Picture Border again to select a line weight. Select Weight. Select the point size line you want to use. Remember that you can preview the border for selecting it by hovering your mouse over that option.

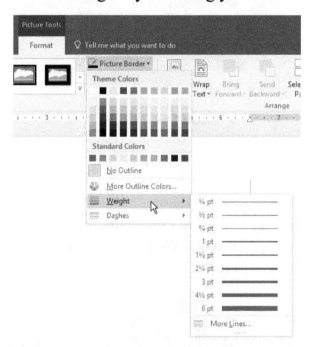

Step 6: Select Picture Border again to select a line style. Select Dashes. Select the line style you want to use.

Removing a Picture's Background

To remove the background from a picture, use the following procedure.

Step 1: Select the picture you want to change.

Step 2: On the Picture Tools tab of the Ribbon, select Remove Background.

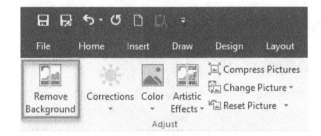

Word displays the Background Removal tab.

Step 3: You can drag the marquee to adjust the picture, if necessary.

Step 4: Select Keep Changes to accept Word's automatic background removal. Or use the Mark Areas to Keep or Mark Areas to Remove tools to refine the background removal. When you have finished, select Keep Changes. Or select Discard All Changes to return to the original picture.

Adding Artistic Effects

To add artistic effects to a picture, use the following procedure.

Step 1: Select the picture you want to change.

Step 2: On the Picture Tools tab of the Ribbon, select Artistic Effects.

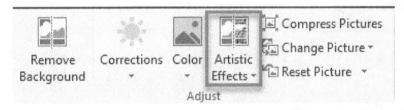

Step 3: Select the effect you would like to apply.

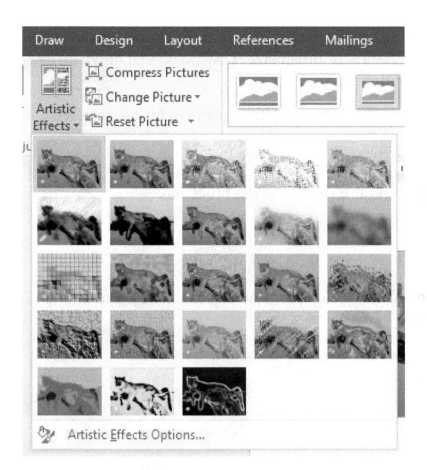

Step 1: Select Artistic Effects Options from the Artistic Effects gallery.

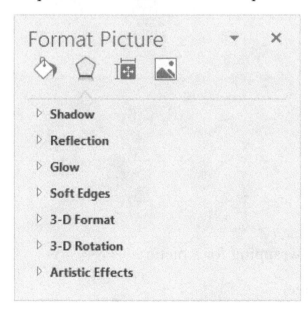

Step 2: Select the Artistic Effect from the drop-down list.

Step 3: Depending on which effect you select, there are different options to adjust, such as transparency, pressure, or brush size. Use the up and down arrows or enter the amounts for each option.

Step 4: Select the Reset button to return to the default settings for the selected option.

Positioning Pictures and Wrapping Text

To set the positioning for a picture, use the following procedure.

Step 1: Select the picture you want to change.

Step 2: On the Picture Tools tab of the Ribbon, select Position.

Step 3: Select the Position that you want to use. You can use the Layout dialog box to refine it later, if needed.

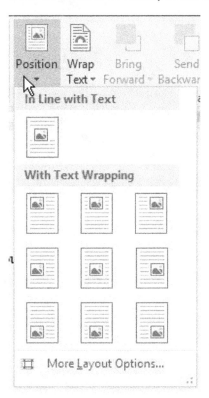

Use the following procedure to set the text wrapping for a picture.

Step 1: Select the picture you want to change.

Step 2: On the Picture Tools tab of the Ribbon, select Wrap Text.

Step 3: Select the wrapping option that you want to use. You can use the Layout dialog box to refine it later, if needed.

The purpose of a newsletter is to provide specialized information to a targeted audience. Newsletters can be a great way to communicate with family and friends on a regular basis.

You can tell stories about your life, your children's activities, your vacations or travel plans, new pets, or whatever you want to tell those closest to you! You can add pictures, too.

You can also find interesting articles and information for your friends to read by accessing the World Wide Web.

Much of the content you put in your newsletter can also be used for your Web site. Microsoft Word offers a simple way to convert your newsletter to a Web publication. So, when you're finished writing your newsletter, convert it to a Web site and post it.

The purpose of a newsletter is to provide specialized information to a targeted audience. Newsletters can be a great way to communicate with family and friends on a regular basis.

You can tell stories about your life, your children's activities, your vacations or travel plans, new pets, or whatever you want to tell those closest to you! You can add pictures, too.

Use the following procedure to use the Layout dialog box.

Step 1: Select the picture you want to change.

Step 2: On the Picture Tools tab of the Ribbon, select EITHER Wrap Text or Position.

Step 3: Select More Layout Options.

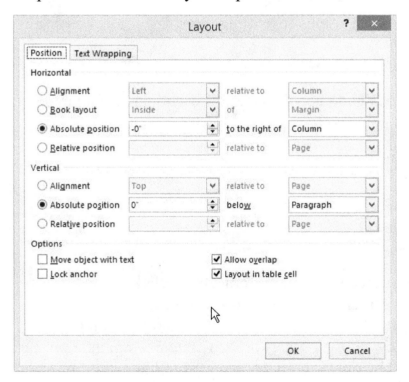

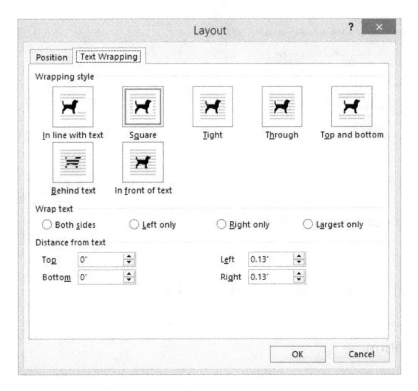

Using the Selection Pane

To open the selection pane, use the following procedure.

Step 1: Select any object on the page to access the Picture Tools Format tab on the Ribbon.

Step 2: Select the Selection Pane tool.

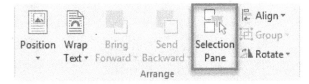

Step 3: In the Selection Pane, you can rename the objects by clicking on an item and entering a new name. You can also use the Send Forward and Send Backward arrow icons to reorder the objects. The Show all and Hide all allow you to hide from the editing view all the objects or show them all again. You can show or hide individual objects by clicking on the eye next to the name for that object.

Chapter 15 – Adding SmartArt

This chapter will show you how to add SmartArt graphics anywhere in your document. You will learn more about the SmartArt Tools tabs, and how to add text to a SmartArt graphic. You will also learn how to move and delete SmartArt graphics. Finally, we will look at the SmartArt Layout options.

Inserting SmartArt

To insert SmartArt, use the following procedure.

Step 1: Select the Insert tab from the Ribbon.

Step 2: Select SmartArt.

Step 3: In the Choose a SmartArt Graphic dialog box, select the category on the left. Then you select the item in the middle. The right shows a preview of the item. Select OK to insert the content.

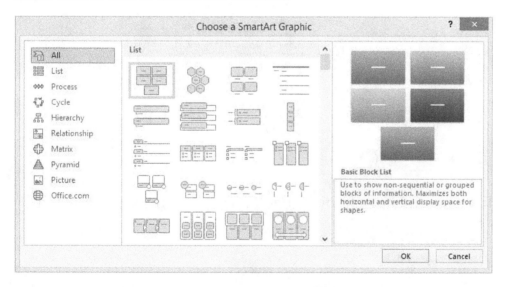

Word inserts the selected SmartArt graphic in the document at the current cursor position.

Adding Text to SmartArt

To add text to a SmartArt graphic by using the Text Pane, use the following procedure.

Step 1: To the left of the SmartArt graphic you inserted, there is a small rectangle with an arrow. Click this arrow to open the Text Pane.

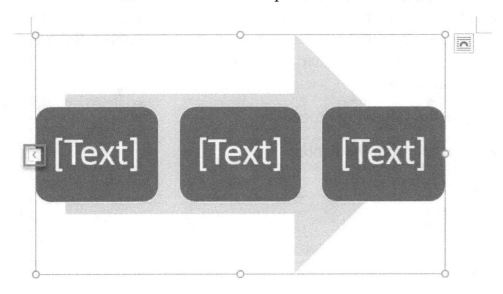

Word opens the Text Pane.

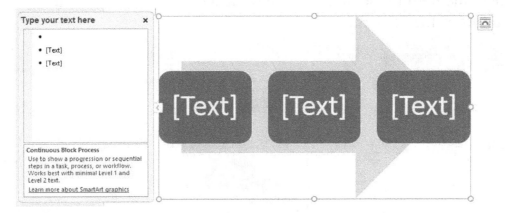

Step 2: Click the first line and begin typing. Each line represents a new item in the graphic.

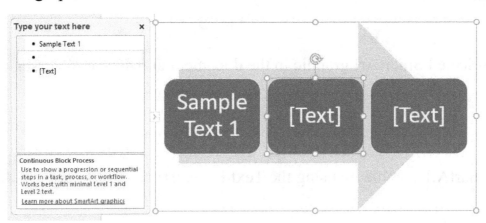

The SmartArt text adjusts to fit the graphic. The more text you enter in each graphic element, the smaller the text will become.

Step 3: When you have finished, click anywhere on the slide, and the Text Pane will close automatically. Or you can click the X in the top right corner.

Using the SmartArt Tools Tabs

Use the following procedure to use the Tools tabs for working with SmartArt.

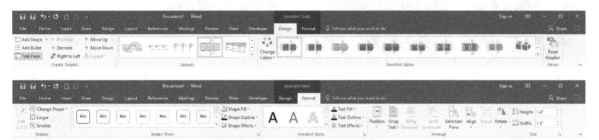

In the Design tab, the Create Graphic tools allow you to customize the SmartArt by adding a shape, adding a bullet point, promoting and demoting or moving a shape right to left, moving a shape up or down and changing the layout. You can also open the Text pane. The change colors option allows you to use the same graphic with different colors.

The reset graphic option removes any changes you have made and returns the selected SmartArt graphic to the default settings. It does not remove your text.

In the Format tab, the tools specific to Smart art allow you to change a selected shape or make it smaller or larger.

Moving and Deleting SmartArt

To move the diagram, use the following procedure.

Step 1: Select the diagram border.

The cursor changes to a cross with four arrows.

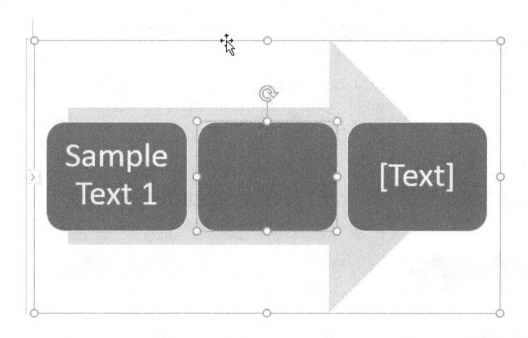

Step 2: Drag the mouse to the desired location. Word displays a small rectangle by the cursor to show an object is being moved. There is a small line showing where the diagram will be moved. Release the mouse to drop the diagram in the new location.

Using SmartArt Layout and Style Options

To change the SmartArt layout, use the following procedure.

Step 1: Select the SmartArt graphic you want to change.

Step 2: On the SmartArt Tools Design tab of the Ribbon, select the down arrow next to the Layout group to see the Layout options.

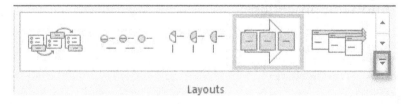

Layouts

Step 3: Select the layout you would like to apply.

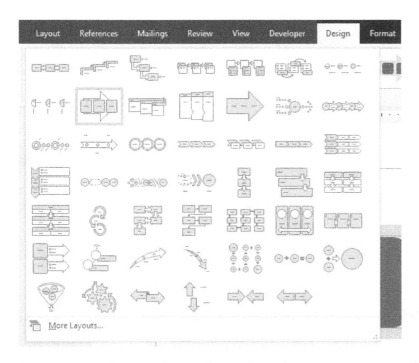

Step 4: To change the style, select the arrow next to the Style group too see the options.

Step 5: Select the style that you would like to apply.

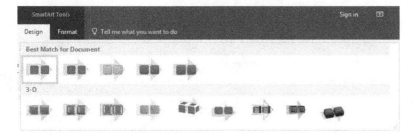

Chapter 16 – Adding Tables

This chapter will explain how to work with tables. You will learn how to add tables to your document and add text to the table. You will also learn about the Table Tools tab. This chapter explains how to modify rows and columns and how to format a table, so that it looks just like you want it to. Finally, you will learn about Quick tables, an easy way to get a table that is already formatted into your document.

Inserting a Table

To insert a table, use the following procedure.

Step 1: Select the Insert tab from the Ribbon.

Step 2: Select Table.

Step 3: Highlight the number of rows and columns that you want to insert.

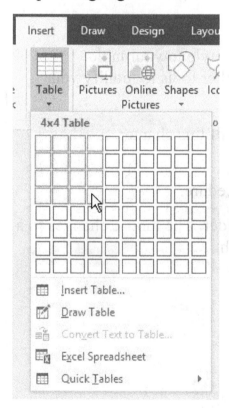

Word inserts the table in the document at the current cursor position. You can also see a preview before you insert the table.

Adding Text to a Table

To add text to a table, use the following procedure.

Step 1: Click the table cell you want to change.

Step 2: Begin typing.

Step 3: To enter text in another cell, click that cell.

Sample Header 1	Sample Header 2	Sample Header 3
Sample Text 1	Sample Text 2	Sample Text 3

About the Table Tools Tabs

The following diagrams show the Tools tabs for working with tables.

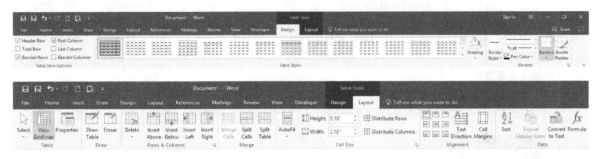

The Design tab has several options to help you apply style to your table, including borders and shading. The Layout tab has other tools to modify your table.

We will investigate many of these options in the rest of this chapter.

Altering Rows and Columns

To modify rows and columns, use the following procedure.

Step 1: Hover your mouse over a row or column divider. The mouse changes to a divider with arrows pointing to the left and to the right.

Sample Header 1	Sample Header 2	Sample Header 3
Sample Text 1	Sample Text 2	Sample Text 3

Step 2: Drag the column to the new size.

Sample Header 1	Sample Header 2	Sample Header 3
Sample Text 1	Sample Text 2	Sample Text 3

To insert a row, use the following procedure.

Step 1: Select the row below where you want the new row to appear.

Step 2: Make sure that the Table Tools/Layout tab is selected.

Step 3: Select Insert Above.

Word inserts the new row.

Sample Header 1	Sample Header 2	Sample Header 3
Sample Text 1	Sample Text 2	Sample Text 3

To delete a column, use the following procedure.

Step 1: Select the column you want to delete.

Step 2: Make sure that the Table Tools/Layout tab is selected.

Step 3: Select Delete.

Step 4: Select Delete Columns.

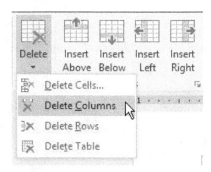

Applying a Table Style

To format a table, use the following procedure.

Step 1: Select the table you want to format.

Step 2: Use the Table Style options to add special formatting to the Header Row, Total Row (last row), First Column, or Last Column. The Banded Rows and Banded Columns alternate the shading.

Step 3: Select a Table style to create a new look for the table. You can see a preview by hovering the mouse over the option before selecting it.

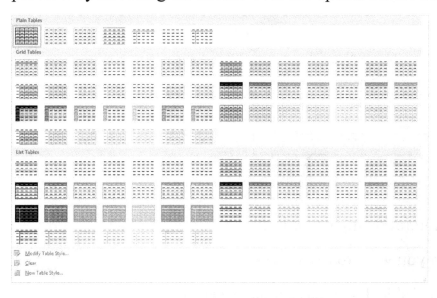

About Quick Tables

To insert a Quick Table, use the following procedure.

Step 1: Select the Insert tab from the Ribbon.

Step 2: Select Table.

Step 3: Select Quick Tables.

Step 4: Select the table you want to insert.

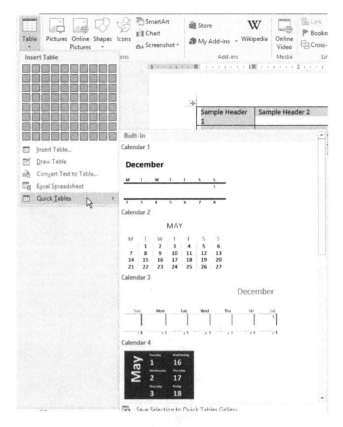

To save a table as a Quick Table selection, use the following procedure.

Step 1: Highlight the table that you have inserted and customized.

Step 2: Select the Insert tab from the Ribbon.

Step 3: Select Table.

Step 4: Select Quick Tables.

Step 5: Select Save Selection to Quick Tables Gallery.

Word displays the Create New Building Block dialog box.

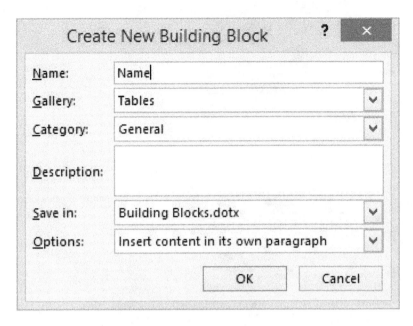

Step 6: Enter a name for the table or leave the default heading.

Step 7: Select OK to save the table.

Chapter 17 – Inserting Special Objects

This chapter will explain how to work with other objects to enhance your documents even further. You will learn how to add a cover page and text boxes to your document. You will also learn about the new features in 2016 to insert an app or online media. Finally, you will learn how to insert the data from a database, using Query Options and Table AutoFormat options.

Adding a Cover Page

To insert a cover page, use the following procedure.

Step 1: Select the Insert tab from the Ribbon.

Step 2: Select Cover Page.

Step 3: Select an option from the Cover Page gallery.

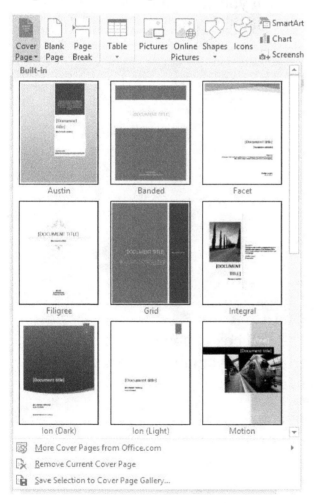

Word inserts the cover page.

Step 4: For each of the elements on the page, click the field and enter the new text. For example, in the above illustration, when you click anywhere on [Type the document title], the entire field is selected. Begin typing to enter the Title.

Inserting a Text Box

To insert a text box, use the following procedure.

Step 1: Place your cursor where you want the text box to appear in the document. Some built-in styles appear to the left or right. However, all text boxes have an anchor somewhere in the text of the document.

Step 2: Select the Insert tab from the Ribbon.

Step 3: Select Text Box.

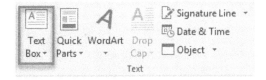

Step 4: Select one of the text box gallery objects, or select Draw Text box.

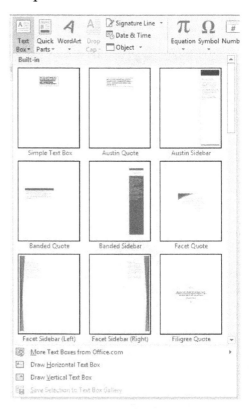

Word inserts the text box. If you selected Draw Text Box, draw the text box just like a shape.

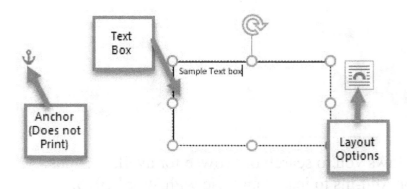

Step 5: Enter your text.

Inserting an Add-in

Apps for Office has been replaced with My Add-ins in Microsoft Word 2016.

To get an Add-in for Office 2016, use the following procedure.

Step 1: Select the Insert tab from the Ribbon.

Step 2: Select My Ad-ins.

Step 3: Select See All.

Step 4: Click Office Store.

The Insert App dialog box allows you to search or browse for available apps. You can search or browse for Add-ins to install and use with Word 2016.

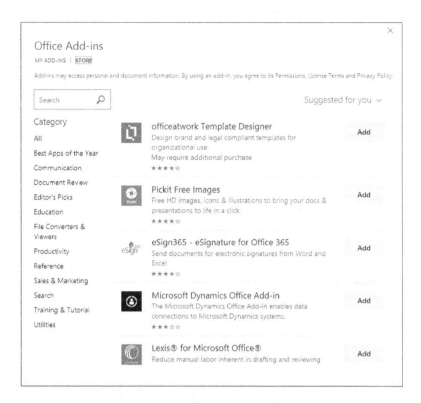

Step 5: Click Add to import the Add-in to your version of Word 2016. The Add-ins will be installed to the Add-in ribbon.

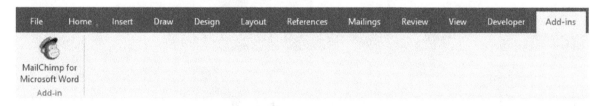

Inserting Online Media

To insert online media, use the following procedure.

Step 1: Select the Insert tab from the Ribbon.

Step 2: Select Online Video.

Step 3: In the Insert Video dialog box, enter a search term in the Bing Video Search field. Or enter the Video Embed Code from a web site.

Step 4: If you entered a search term, Word displays videos that match your search term. Select the video that you want to use, and select Insert.

Step 5: Word inserts the video into the document.

Inserting a Database

To insert data from a database into the document, use the following procedure.

Step 1: Select the Insert tab from the Ribbon.

Step 2: Select Insert Database.

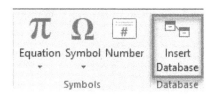

Step 3: In the Database dialog box, select Get Data.

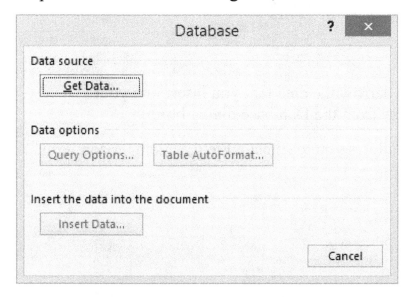

Step 4: Navigate to the location of the database file you want to use. Highlight and select Open.

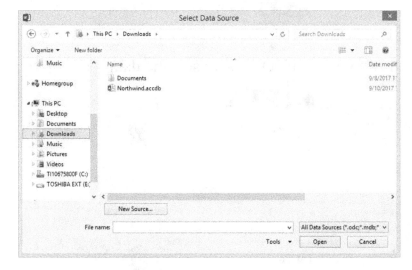

Step 5: If your database includes more than one table, the Select Table dialog box is displayed. Select the table that you want to use and select OK.

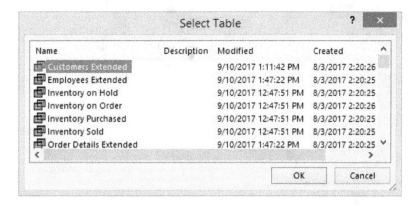

Step 6: You can use a query to narrow the data that you insert into your document. Select Query Options from the Database dialog box.

Step 7: The Filter Records tab allows you to select a Field, a Comparison term, and the details for the filter.

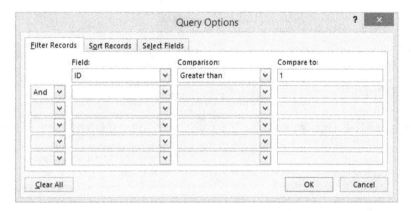

Step 8: The Sort Records tab allows you to select how to sort the data.

Step 9: The Select Fields tab allows you to choose which fields to show.

Step 10: Select OK when you have finished setting your Query Options.

Step 11: You can use an automatic format for the table Word will insert. Select Table AutoFormat. You can choose a Format and see a Preview. The checkboxes allow you to choose which formats and special formats to apply to your table. Select OK when you have finished.

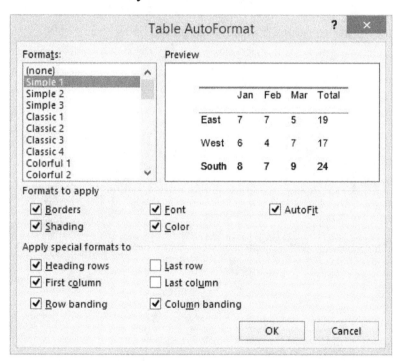

Step 12: The Insert Data dialog box allows you to choose all records or from a select set. To choose a select set, enter the Start and End records. You can choose to insert data as a Word field. Select OK.

Step 13: Word inserts the data as a table.

File As	Contact Name	ID	Company	Last Name	First Name	E-mail Address	Job Title	Business Phone	Ho Ph.
Andersen, Elizabeth	Elizabeth Andersen	8	Company H	Andersen	Elizabeth		Purchasing Representative	(123)555-0100	
Autier Miconi, Catherine	Catherine Autier Miconi	18	Company R	Autier Miconi	Catherine		Purchasing Representative	(123)555-0100	
Axen, Thomas	Thomas Axen	3	Company C	Axen	Thomas		Purchasing Representative	(123)555-0100	
Bagel, Jean Philippe	Jean Philippe Bagel	17	Company Q	Bagel	Jean Philippe		Owner	(123)555-0100	
Edwards, John	John Edwards	12	Company L	Edwards	John		Purchasing Manager	(123)555-0100	
Eggerer, Alexander	Alexander Eggerer	19	Company S	Eggerer	Alexander		Accounting Assistant	(123)555-0100	
Entin, Michael	Michael Entin	23	Company W	Entin	Michael		Purchasing Manager	(123)555-0100	
Goldschmidt, Daniel	Daniel Goldschmidt	16	Company P	Goldschmidt	Daniel		Purchasing Representative	(123)555-0100	
Gratacos Solsona, Antonio	Antonio Gratacos Solsona	2	Company B	Gratacos Solsona	Antonio		Owner	(123)555-0100	
Grilo, Carlos	Carlos Grilo	14	Company N	Grilo	Carlos		Purchasing Representative	(123)555-0100	

This chapter explains how to use Word's reference tools. First, we will discuss how to add a caption to an illustration. You can add an automatically generated table of contents. Word makes it easy to add footnotes, endnotes, and other citations. Once you have added references, the Manage Sources tool helps you to keep track of those sources, which can be especially helpful in a long document or when sharing references across multiple documents. This chapter will explain how to insert a bibliography. We will end with a discussion on creating an index.

Inserting a Caption

To add a caption, use the following procedure.

Step 1: Select the References tab from the Ribbon.

Step 2: Select Insert Caption.

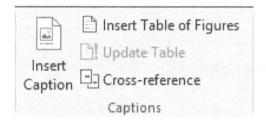

Step 3: In the Caption dialog box, enter the text for your Caption. The Label and the Numbering are shown by default.

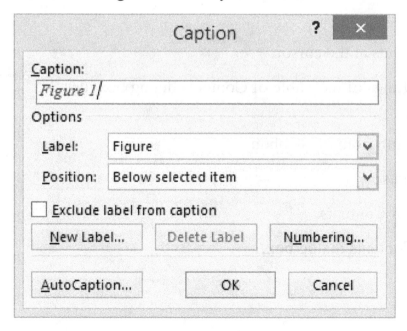

Step 4: To change the Label, select a new option from the Label drop down list. You can also select New Label to create a custom label. Just enter the text and

select OK, and it will be added to the drop-down list of options for Labels. You can select Delete Label to remove it from the list.

Step 5: You can select a new Position for the caption by selecting an item from the drop-down list.

Step 6: Select Numbering to choose the format for the caption number.

Step 7: Select OK to add your Caption.

Adding a Table of Contents

To add a table of contents, use the following procedure.

Step 1: Place your cursor in the document where you want the table of contents to appear.

Step 2: Select the References tab from the Ribbon.

Step 3: Select Table of Contents.

Step 4: Select one of the built-in Table of Contents styles.

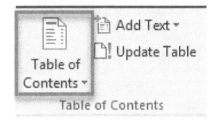

Word inserts the table of contents at the cursor.

To review the options and settings of the Table of Contents dialog box, use the following procedure.

Step 1: Select the References tab from the Ribbon.

Step 2: Select Table of Contents.

Step 3: Select Custom Table of Contents.

Word displays the Table of Contents dialog box.

Step 4: To insert a customized table of contents, complete the Table of Contents dialog box.

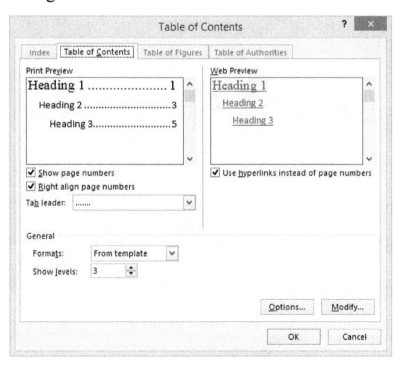

- You can see a preview of your selections for box print and web distribution.

- You can select whether to show the page numbers, and what kind of tab leader to use between the headings and page numbers.

- You can select what kind of style to use and how many heading levels to include.

- Select Options to open the Options dialog box. Here you can indicate which paragraph styles to include in the table of contents at each TOC level. Select OK when you have finished.

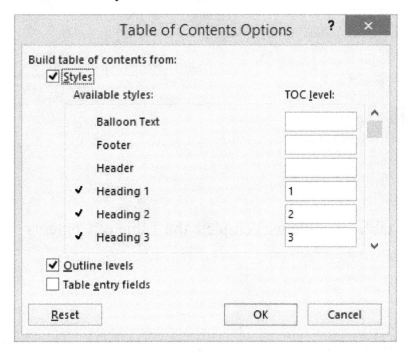

- Select Modify to change the appearance of the table of contents entries. Select OK when you have finished.

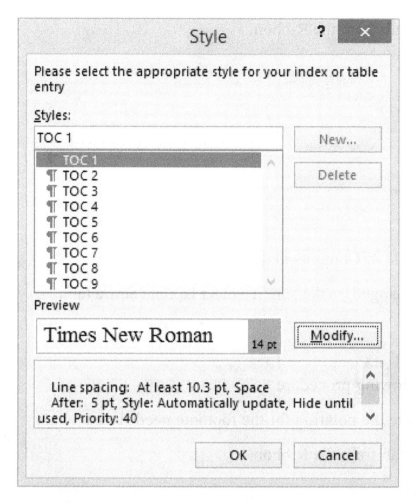

To update a table of contents, use the following procedure.

Step 1: Click anywhere on the table.

Step 2: Select the References tab from the Ribbon.

Step 3: Select Update Table.

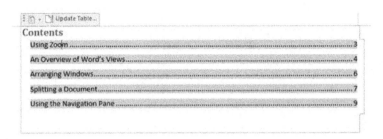

Word displays the Update Table of Contents dialog box.

Step 4: If the headings have changed, make sure to select Update entire table.

Step 5: Select OK.

Adding Footnotes, Endnotes, and Citations

To add a footnote, use the following procedure.

Step 1: Place your cursor where the notation for the footnote needs to go.

Step 2: Select the References tab from the Ribbon.

Step 3: Select Insert Footnote.

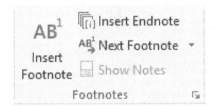

Word inserts a number at the cursor location. It is already formatted as a superscript font. Word also inserts a line and the matching number at the bottom of the page. It also places the cursor so that you can type the footnote text.

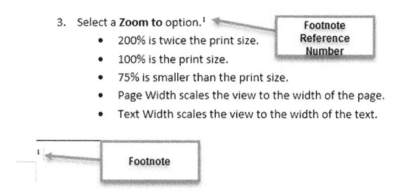

Step 4: Begin typing the footnote text.

To add an endnote, use the following procedure.

Step 1: Place your cursor where the notation for the endnote needs to go.

Step 2: Select the References tab from the Ribbon.

Step 3: Select Insert Endnote.

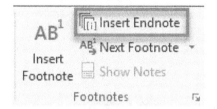

Word inserts a number at the cursor location. It is already formatted as a superscript font. Word also inserts a line and the matching number at the end of the document. It also places the cursor so that you can type the endnote text.

---

[i] Sample End Note

To insert a citation, use the following procedure.

Step 1: Place your cursor in the paragraph that needs to be referenced.

Step 2: Select the References tab from the Ribbon.

Step 3: Select Insert Citation.

Step 4: To enter a placeholder, select Add New Placeholder.

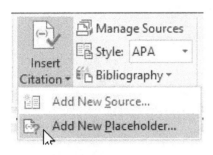

Step 5: In the Placeholder Name dialog box, enter a tag name to help you remember the source. The name cannot include any spaces or special characters. Select OK.

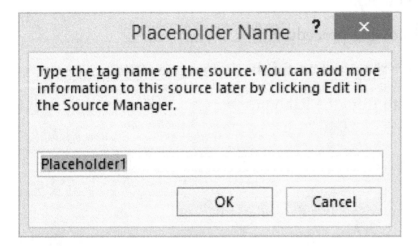

Step 6: To enter a source, select Add New Source from the Insert Citation command on the Ribbon. In the Create Source dialog box, enter the bibliography information. Select OK.

Managing Sources

To use the Source Manager, use the following procedure.

Step 1: Select the References tab from the Ribbon.

Step 2: Select Manage Sources.

Citations & Bibliography

Word opens the Source Manager dialog box.

Explain the different areas of the Source Manager dialog box.

The Search fields allow you to narrow the list in the Sources available in area. Select the Sort By option from the drop-down list and begin typing the name in the Search field. Word automatically narrows the list displayed to any matching options.

Word includes a master citation list for your computer. To open the list for another document, select Browse. Navigate to the location of the source file (in XML format), highlight it, and select Open.

The left list includes the citations in the selected citation file or the Master List. You can Copy, Delete, Edit, or create a New citation for the current list or the Master List.

- To copy a citation from one list to the other, highlight the citation and select Copy.

- To delete a citation in either list, highlight it and select Delete.

- To edit a citation, highlight it and select Edit. Word opens the Edit Source dialog box, which includes the same information as the Create Source dialog box from the previous topic.

- To create a new citation, select New. Word opens the Create Source dialog box.

The bottom area of the source manager dialog box displays the preview for how the currently highlighted citation will look in the bibliography.

Select Close when you have finished working with the sources.

To insert a bibliography, use the following procedure.

Step 1: Place your cursor in the location where you want to add the bibliography.

Step 2: Select the References tab from the Ribbon.

Step3: Select Bibliography.

Step 4: Select the desired style of Bibliography from the Bibliography gallery.

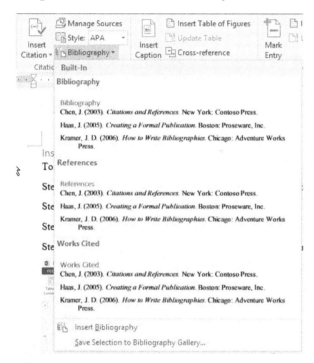

Word inserts the bibliography, including tools to change the style of the bibliography and update the citations.

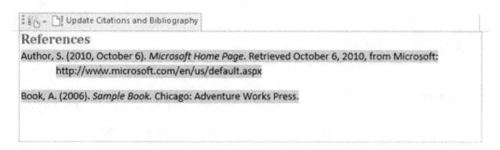

Creating an Index

To mark an index entry, use the following procedure.

Step 1: If desired, highlight the text you want to include in the index from your document. Or simply place your cursor in the line where the index entry should point.

Step 2: Select the References tab from the Ribbon.

Step 3: Select Mark Entry.

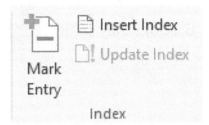

Word opens the Mark Index Entry dialog box. If you highlighted text, that text appears as the Main Entry.

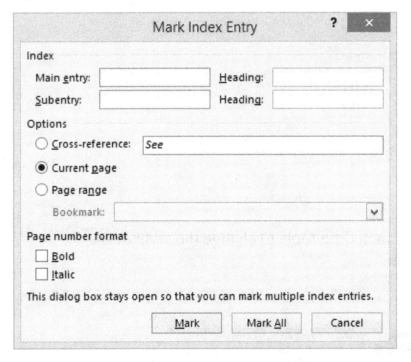

Step 4: Enter the text for the Main entry and the subentry (if desired).

Step 5: Select whether to refer to a cross reference (enter the text), the current page, or a page range (select a bookmark from the drop-down list).

Step 6: Select any special formatting for the page number.

Step 7: Select Mark or Mark All.

Word inserts an index marker, which you can see if you have the paragraph markers visible. Note that these markers are not visible in the printed document.

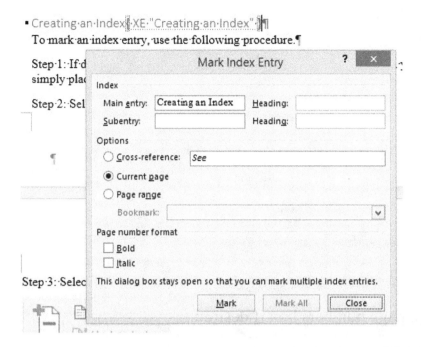

Step 8: The dialog box will stay open if you need it. Select Close when you have finished marking index entries.

To insert an index, use the following procedure.

Step 1: After you have marked at least some of your index entries, place your cursor where you want the index to appear.

Step 2: Select the References tab from the Ribbon.

Step 3: Select Insert Index.

Word opens the Index dialog box.

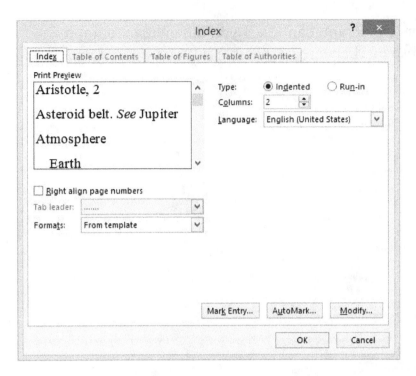

Step 4: Check the Right align page numbers box if desired.

Step 5: Select the Type as Indented or Run-in.

Step 6: Select the number of Columns.

Step 7: Select the Language.

Step 8: Select OK.

Word inserts the index at the current location.

This chapter explains how to use Word's advanced research tools. The dictionary, thesaurus and Word count tools are proofing tools that each opens a separate task pane or dialog box. The Translation tools allow you to translate a document or selected text. You can also use the mini translator to obtain a quick translation that is only temporarily visible. Finally, we will discuss the language tools to help you set your proofing language and other language preferences.

Using Smart Lookup, Thesaurus and Word Count

The Dictionary feature in Microsoft Word 2016 has been replaced with Smart Lookup, a more advanced reviewing tool. This tool is not on the Review ribbon by default, but you can still access the tool. To use Smart Lookup, use the following procedure.

Step 1: Highlight the word you would like to research.

Step 2: Right-click the word and select Smart Lookup from the pop-up menu.

Step 3: The Smart Lookup window has two options: Explore and Define.

Explore provides a definition of the term, explores Wikipedia, and performs a web search. Define focuses primarily on the definition. The pronunciation is provided in text and as an audio file.

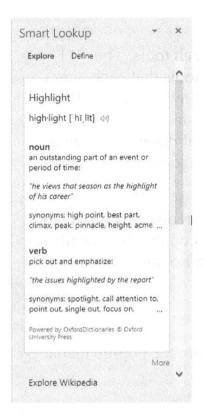

Add Smart Lookup Icon to the Review ribbon

Step 1: Click the File ribbon.

Step 2: Select Options.

Step 3: Select the Customize Ribbon option.

Step 4: Select the Review Ribbon in the Tabs pane.

Step 5: Click New Group to add a new group to the ribbon.

Step 6: Click Rename to give the new group a name.

Step 7: Select All Commands from the Choose Command From drop-down.

Step 8: Scroll down and select Smart Lookup and click Add to add the command to the ribbon.

Step 9: Click Ok.

To use the thesaurus, use the following procedure.

Step 1: Highlight the word you would like to research.

Step 2: Select the Review tab.

Step 3: Select Thesaurus.

Word displays the Thesaurus task pane (with the dictionary that you have previously installed as an app) with the results for the word you selected displayed.

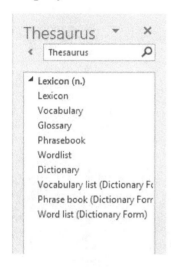

Step 4: To replace the selected word in the document with one of the words listed in the Thesaurus pane, right click the word you want to use and select Insert from the context menu.

Access the Word Count tool

Step 1: Click the Review tab.

Step 2: Click the Word Count icon in the Proofing menu.

In the Word Count dialog box. Note that the statistics show for selected text (if applicable) or the entire document.

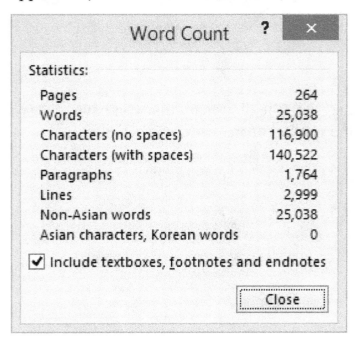

Using Translation Tools

To use the Translation tools, use the following procedure.

Step 1: Select the Review tab.

Step 2: Select Translate expanded menu.

Step 3: Select Choose Translation Language.

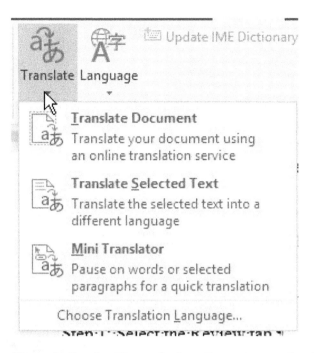

Step 4: In the Translation Language Options dialog box, you will select the language to use in the Mini Translator first. Below, you set the To and from languages for document translation. For all three options, select the language you want to use from the drop-down list.

Step 5: Select OK.

Step 6: Select Translate Document to translate the entire document. You may see a message that reads "To translate your document, text will be sent over the Internet in a secured format to Microsoft or a third-party translation service. Do you want to proceed?" Click Yes or No. You can place a check in the box next to "Do not show again" to prevent this message from popping up every time you request a translation.

The text opens in Internet Explorer using an online translation service.

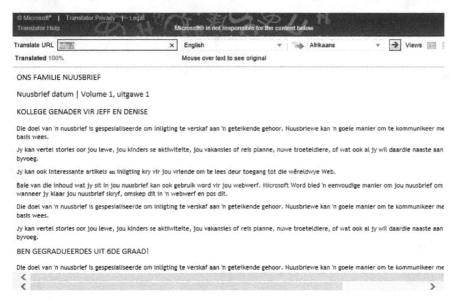

Step 7: Select Translate Selected text to see a translation of just a few words. The Research task pane opens with the current translation settings and your translated text.

To change the language, use the following procedure.

Step 1: Select the Review tab.

Step 2: Select Language. Select Set Proofing Language.

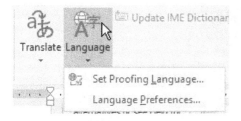

Step 3: Select the language you want to use from the Language dialog box and select OK.

To set Language Preferences in the Word Options dialog box, use the following procedure.

Step 1: Select Language from Review tab on the Ribbon.

Step 2: Select Language Preferences.

Step 3: In the Word Options dialog box, select the Editing Language you want to use for dictionaries, spell check and sorting.

Word Options

General
Display
Proofing
Save
Typography
Language
Ease of Access
Advanced
Customize Ribbon
Quick Access Toolbar
Add-ins
Trust Center

Set the Office Language Preferences.

Choose Editing Languages

Add additional languages to edit your documents. The editing languages set language-specific features, including
dictionaries, grammar checking, and sorting

Editing Language	Keyboard Layout	Proofing (Spelling, Grammar...)
English (United States) <default>	Enabled	Installed
Chinese (Taiwan)	Not enabled	Not installed

Remove
Set as Default

[Add additional editing languages]   ▼   Add

☐ Let me know when I should download additional proofing tools.

Choose Display Language

Set the language priority order for the buttons, tabs and Help ⓘ

Display Language		Help Language	
1. Match Microsoft Windows <default>		1. Match Display Language <default>	
2. English		2. English	

Set as Default   Set as Default

How do I get more Display and Help languages from Office.com?

OK   Cancel

Step 4: Select OK.

This chapter will explain how to review a document. You can add, reply to, or review comments. Comments are separate from the main text of the document. Track changes, on the other hand, allow you to make changes directly to the document in such a way that other reviewers can see your changes. Then you can review those changes and decide whether to keep them or not. Finally, this chapter explains how to compare different documents.

Adding a Comment

To add a comment, use the following procedure.

Step 1: Place the cursor where you want to mark a comment or highlight the portion of text on which you want to comment.

Step 2: Select the Review tab from the Ribbon.

Step 3: Select New Comment.

Word inserts a comment bubble with the Comments window open.

Step 4: Enter the comment text.

The Comment window closes when you click somewhere else in the document. You can also close it by clicking the X at the top right corner. To open it again, click the Comments bubble near the right margin.

To reply to a comment, use the following procedure.

Step 1: In the Comments window, click the Reply icon.

Step 2: Enter your text.

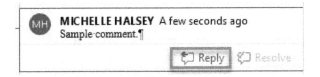

To review comments, use the following procedure.

Step 1: Select the Review tab from the Ribbon.

Step 2: Select Show Comments.

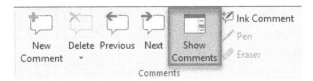

Step 3: Use the Next and Previous tools to move from one comment to the next.

Step 4: Review the comments in the Markup area.

To delete a comment, use the following procedure.

Step 1: Place your cursor anywhere in the selection for the comment you want to delete.

Step 2: Select the Review tab from the Ribbon.

Step 3: Select Delete.

Tracking Changes

Use the following procedure to track changes.

Step 1: Select the Review tab from the Ribbon.

Step 2: Select Track Changes.

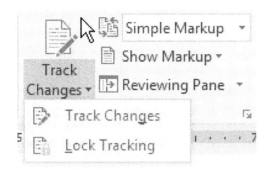

Step 3: Make edits to the document. Word places a line next to any area with changes. It marks insertions, deletions, moves, and formatting changes according to the settings in the Change Tracking Options dialog box. However, you may not see all markups, depending on your settings.

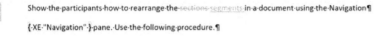

Show·the·participants·how·to·rearrange·the·sections·segments·in·a·document·using·the·Navigation¶

{·XE·"Navigation"·}·pane.··Use·the·following·procedure.¶

Reviewing Changes

To open the Reviewing Pane, use the following procedure.

Step 1: Select the Review tab from the Ribbon.

Step 2: Select Reviewing Pane.

Step 3: Select the orientation you would like to use for the Reviewing pane.

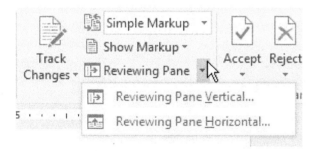

Word displays the Revisions Pane. The different authors who have made changes are indicated with a description of the change. When you click an item in the Revisions Pane, Word automatically scrolls to the corresponding location in the document.

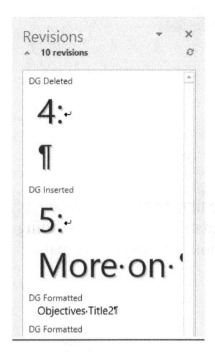

Select Previous or Next to move to another tracked change. Select Accept or Reject to accept or reject the current change.

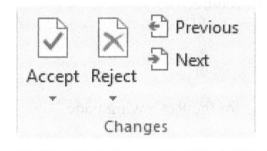

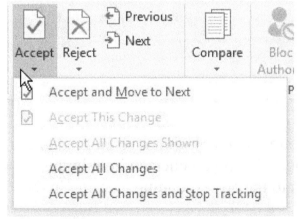

Comparing and Combining Documents

To compare documents, use the following procedure.

Step 1: Select the Review tab from the Ribbon.

Step 2: Select the Compare expanded options.

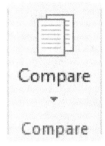

Step 3: Select Compare from the drop-down menu.

Word opens the Compare Documents dialog box to determine which documents to use.

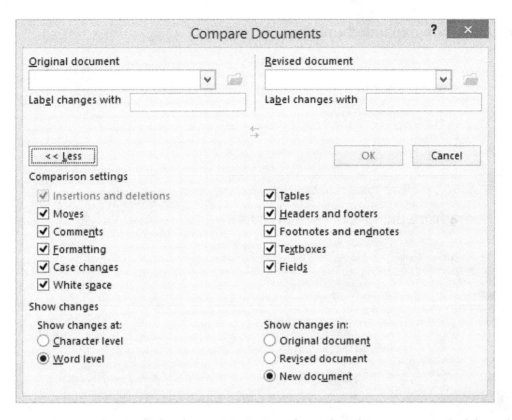

Step 3: Under Original Document, select the document considered the original from the drop-down list. If the document is not listed, select the folder icon to navigate to the document and select Open. To label this document's changes, enter the label in Label change with field.

Step 4: Under Revised Document, select the document considered the revised document from the drop-down list. If the document is not listed, select the folder icon to navigate to the document and select Open. To label this document's changes, enter the label in Label change with field.

Step 5: Select More to indicate which Comparison settings you want to mark. You can check or clear any of the boxes to control which items are compared. You can select whether to show changes at a character level or a word level. You can show changes in the Original, the Revised document, or a New document.

Step 6: Select OK to compare the documents.

Word compares the document. Note the Revisions pane, the Comparison document, the original document, and the revised document open in different panes.

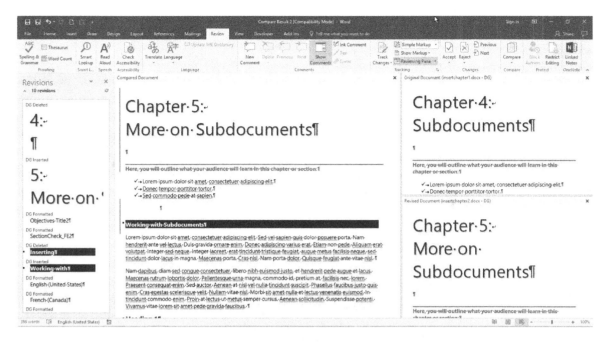

You will need to save the comparison document if you want to keep it.

To combine documents, use the following procedure.

Step 1: Select the Review tab from the Ribbon.

Step 2: Select the Compare expanded options.

Step 3: Select Combine from the drop-down menu.

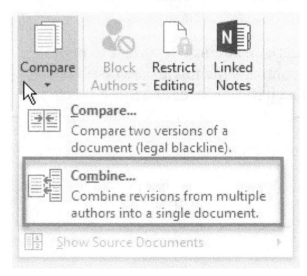

Word opens the Combine Documents dialog box to determine which documents to use.

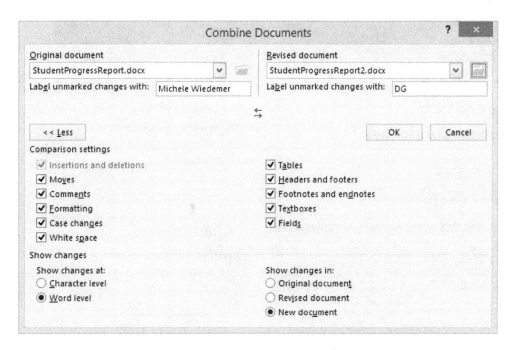

Step 3: Under Original Document, select the document considered the original from the drop-down list. If the document is not listed, select the folder icon to navigate to the document and select Open. To label this document's changes, enter the label in Label change with field.

Step 4: Under Revised Document, select the document considered the revised document from the drop-down list. If the document is not listed, select the folder icon to navigate to the document and select Open. To label this document's changes, enter the label in Label change with field.

Step 5: Select More to indicate which Comparison settings you want to mark. You can check or clear any of the boxes to control which items are combined. You can select whether to show changes at a character level or a word level. You can show changes in the Original, the Revised document, or a New document.

Step 6: Select OK to combine the documents.

Word combines the document. Note the Revisions pane, the combined document, the original document, and the revised document open in different panes.

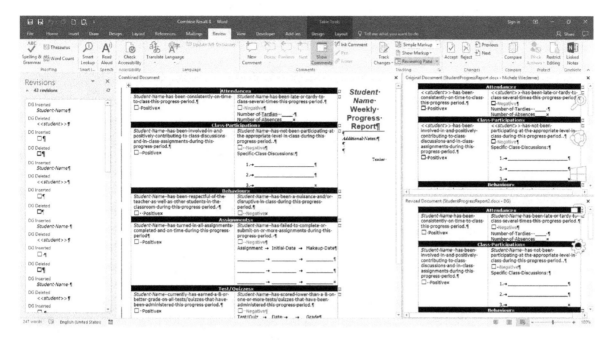

You will need to save the comparison document if you want to keep it.

The Combine documents process is the same procedure. If there are changes, Word can only store one set.

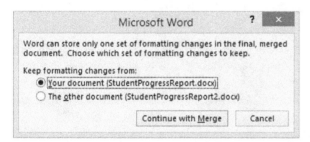

You will need to save the combined document if you want to keep it.

Chapter 21 – Customizing Word

In this chapter, you'll understand how to control the environment where you create documents, as well as specialized information about those documents. We'll start with customizing your Word options to make the environment perfectly suited to your use. Then, you'll learn how to protect a document. You'll also learn how to check for issues when working with others who are using earlier versions of Word. This module also explains how to manage different versions of a document. Finally, you'll learn the basics of the document properties and information.

Setting Word Options

To review the options for customizing Word, use the following procedure.

Step 1: Select the File tab from the Ribbon to open the Backstage view.

Step 2: Select the Options tab on the left.

Here is the General tab in the Word Options dialog box. The General tab allows you to change the user interface options. You can enter your name and initials to personalize your copy of Word. You can also control Start up options.

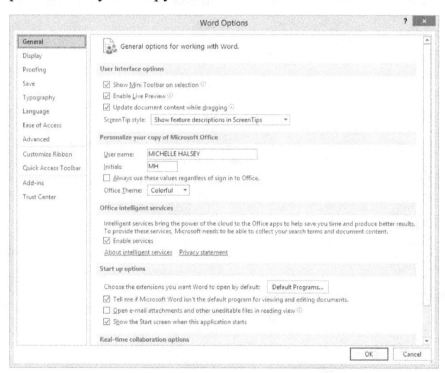

Here is the Display tab in the Word Options dialog box. The Display tab controls the Page display, the formatting marks, and the printing options.

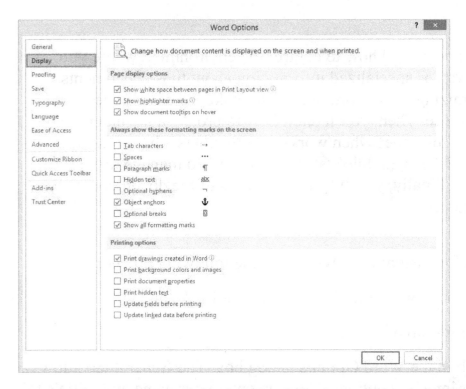

Here is the Proofing tab in the Word Options dialog box. The Proofing tab allows you to control how Autocorrect works for spelling, grammar, and formatting.

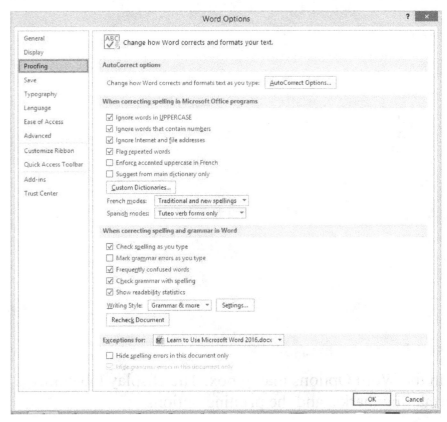

Here is the Save tab in the Word Options dialog box. The Save tab allows you to control how documents are saved.

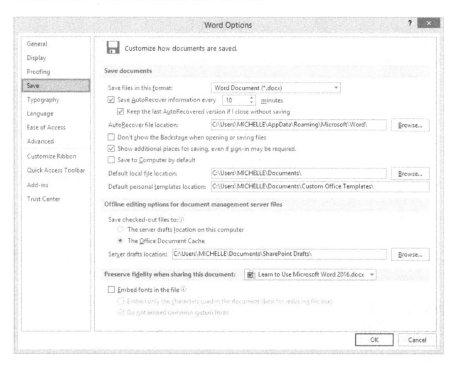

Here is the Typography tab in the Word Options dialog box. The Typography tab allows you to control character sets, character spacing, and kerning.

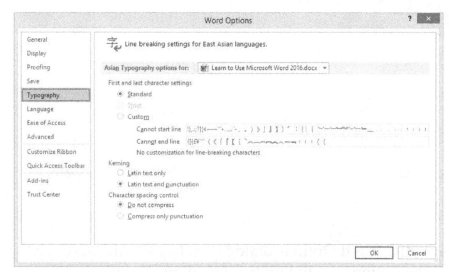

Here is the Language tab in the Word Options dialog box. The Language tab allows you to choose a language for use while editing your documents, which controls the spell checker and grammar. You can also change the language of the help files.

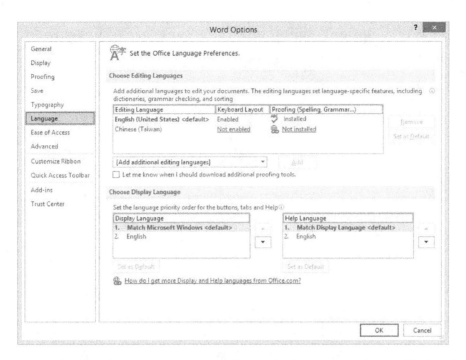

Here is the Ease of Access tab in the Word options dialog box. In the Ease of Access tab, you can set accessibility options.

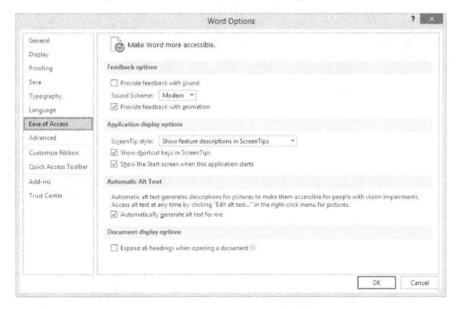

Here is the Advanced tab in the Word Options dialog box. In the Advanced tab, you can change many editing options, including the default paste option.

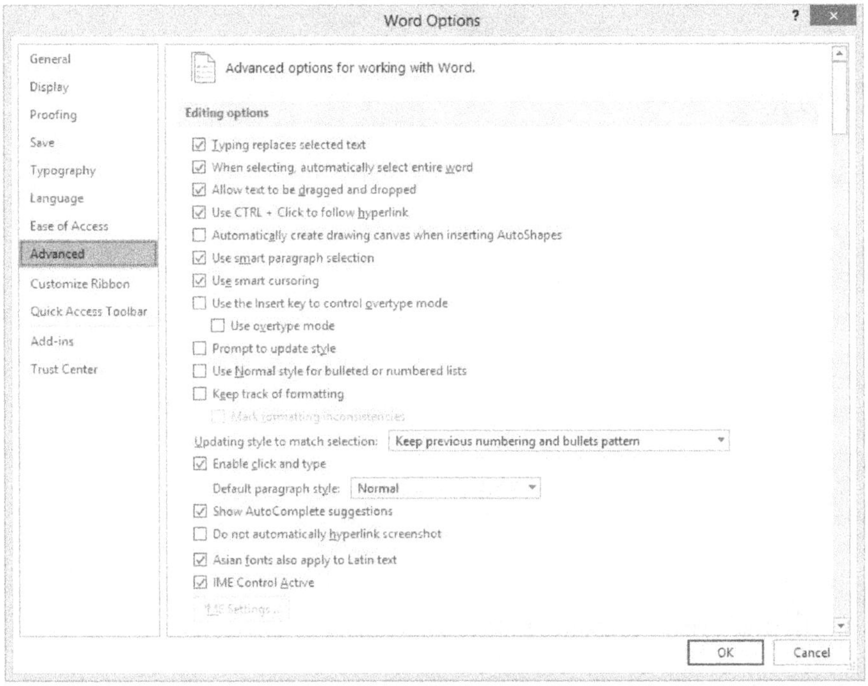

Protecting a Document

Use the following procedure to password protect a document.

Step 1: Select the File tab from the Ribbon to open the Backstage view.

Step 2: Select Info, if it isn't already selected.

Step 3: Select Protect Document.

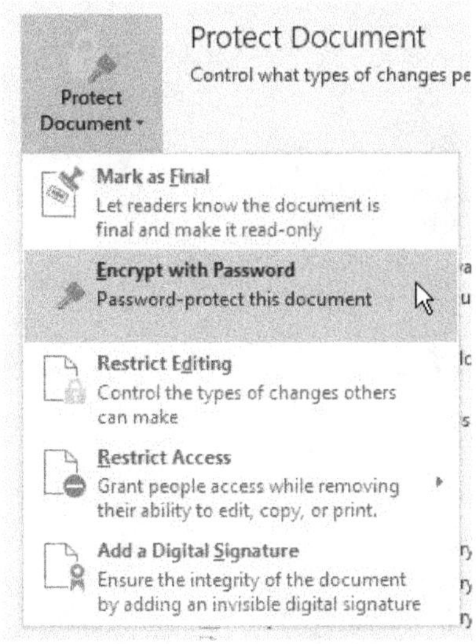

Step 4: Select Encrypt with Password.

Step 5: Enter a password and select OK. Make sure to keep the password safe, because the file cannot be recovered if you lose the password.

To inspect a document for hidden properties or personal information, use the following procedure.

Step 1: Select the File tab from the Ribbon to open the Backstage view.

Step 2: Select Info, if it isn't already selected.

Step 3: Select Check for Issues.

Step 4: Select Inspect Document.

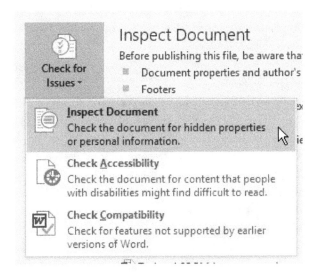

Step 5: In the Document Inspector dialog box, check the boxes for the types of issues you want to check for.

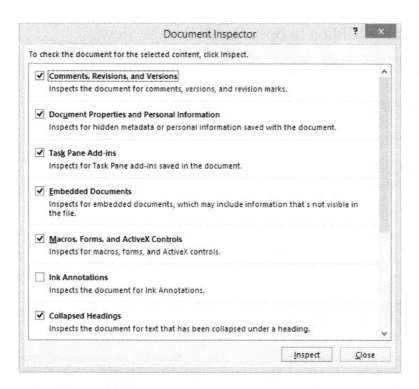

Step 6: Select Inspect.

Step 7: The Document Inspector dialog box indicates any issues with the document. You can select Remove All to remove the features from the document.

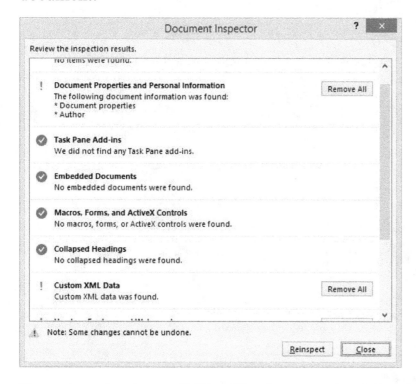

To review the Accessibility Checker, use the following procedure.

Step 1: Select the File tab from the Ribbon to open the Backstage view.

Step 2: Select Info, if it isn't already selected.

Step 3: Select Check Accessibility.

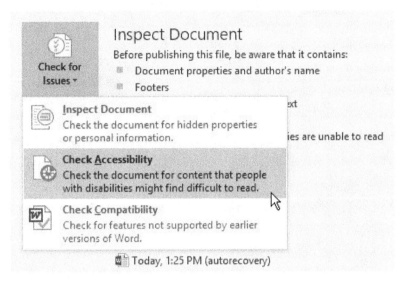

The Accessibility Checker appears on the right side of the document window. The Inspection Results area highlights any problems. You can click on an item in the list to go to that item to correct it. You can also why to fix and how to fix the item.

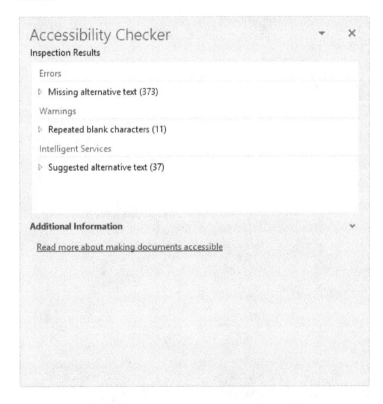

Use the following procedure to review the Compatibility Checker.

Step 1: Select the File tab from the Ribbon to open the Backstage view.

Step 2: Select Info, if it isn't already selected.

Step 3: Select Check Compatibility.

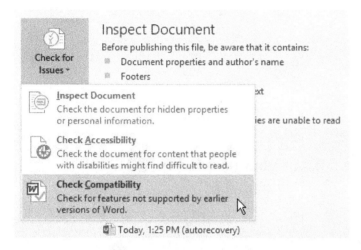

The Microsoft Word Compatibility checker dialog box shows features that are not supported by earlier versions of Word. You can select the version of Word from the Select Versions to Show drop down list.

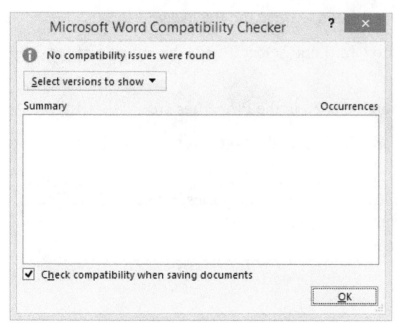

Managing Versions

To use the Manage Versions feature, use the following procedure.

Step 1: Select the File tab from the Ribbon to open the Backstage view.

Step 2: Select Info, if it isn't already selected.

Step 3: Select the Manage Document expanded menu.

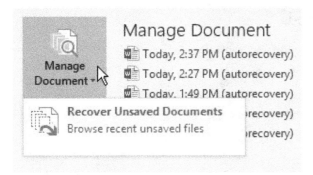

Step 4: Select Recover Unsaved Documents.

Step 5: The Open dialog box displays a list of your unsaved files. Highlight the file and select Open.

Step 6: Make sure you save the file.

Working with Properties

To review the information contained in the Information tab on the Backstage View, use the following procedure.

Step 1: Select the File tab on the Ribbon to open the Backstage View.

Step 2: It should open to the Info tab. If not, select it from the left side of the screen.

Discuss the different information included in the Properties area.

Step 3: Select the Show All Properties link at the bottom to see additional information.

Step 4: You can chance the Title, Tags, Status, Categories, Subject, Hyperlink Base, and Company information.

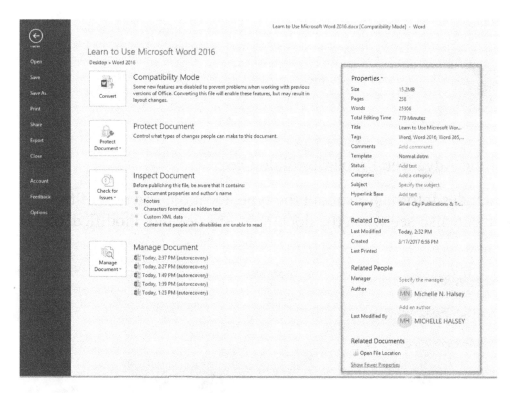

To view or add People information, use the following procedure.

Step 1: To view an author or last modified person, click on the name. You can see the Microsoft Office contact card for that person.

Step 2: To add an author or manager, click on the Specify the Manage or Add an Author field. Enter the name. You can use the icons to verify the information or use the Address book to find the name.

To open the Advanced Properties dialog box, use the following procedure.

Step 1: Select the arrow next to Properties.

Step 2: Select Advanced Properties from the drop-down list.

Total Editing Time   779 Minutes

Discuss the tabs in the Advanced Properties dialog box.

The General tab includes information about the type, location, and size of the document file. You can also see when the document was created, modified, and accessed.

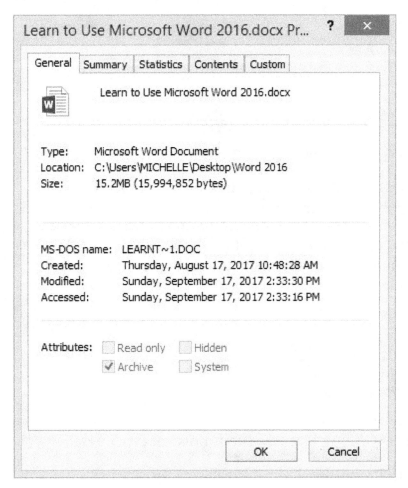

The Summary tab includes details about the document title, subject, author, etc. You can enter information in any of the Summary fields.

The Statistics tab includes information about the document revisions and other statistics.

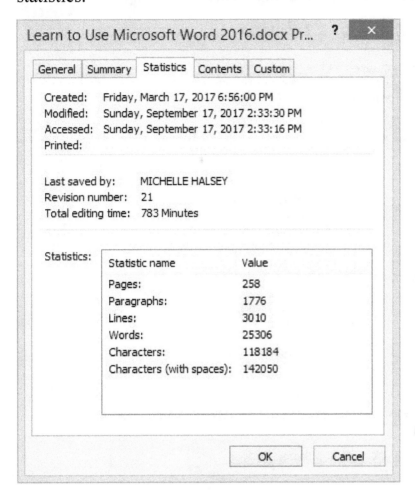

Learn to Use Microsoft Word 2016.docx Pr...	? ✕

General | Summary | **Statistics** | Contents | Custom

Created: Friday, March 17, 2017 6:56:00 PM
Modified: Sunday, September 17, 2017 2:33:30 PM
Accessed: Sunday, September 17, 2017 2:33:16 PM
Printed:

Last saved by: MICHELLE HALSEY
Revision number: 21
Total editing time: 783 Minutes

Statistics:

Statistic name	Value
Pages:	258
Paragraphs:	1776
Lines:	3010
Words:	25306
Characters:	118184
Characters (with spaces):	142050

OK | Cancel

The Contents tab includes information about the document Metadata contained in the file.

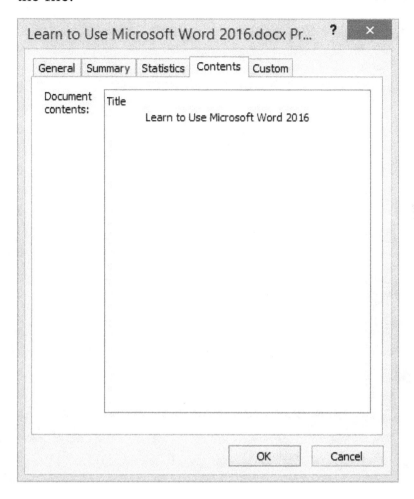

The Custom tab allows you to add custom properties to your document. To add a custom property, complete the following steps.

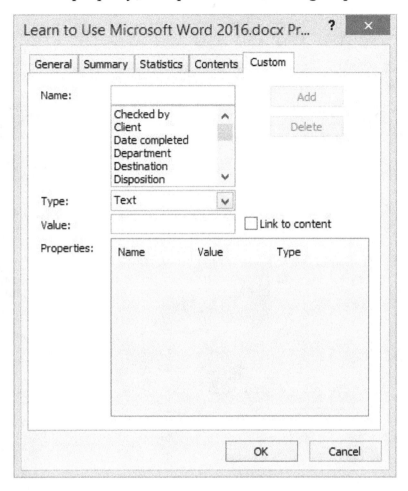

Step 1: Select a Name from the list or enter your own name.

Step 2: Select the Type of information from the drop-down list.

Step 3: Enter the Value for the information.

Step 4: Select Add.

Microsoft Word has many ways to reuse content, such as small snippets of text or even images and whole pages of formatting. This chapter will start by looking at Autotext, which is a type of Quick Part in Word. Then you'll learn how to insert a Quick Part. This module also explains how to create customized building blocks to really help save you time. Finally, you'll learn how to edit a building block.

Saving Selection as Autotext

To create an Autotext entry, use the following procedure.

Step 1: Select the text you want to store.

Step 2: Select the Insert tab from the Ribbon.

Step 3: Select Quick Parts.

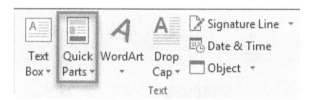

Step 4: Select AutoText from the drop-down list.

Step 5: Select Save Selection to AutoText Gallery.

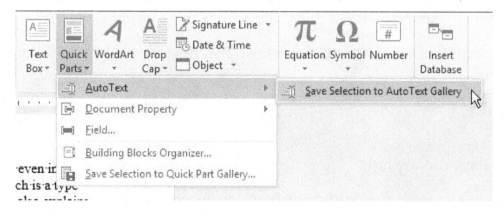

Step 6: In the Create New Building Block dialog box, enter a Name for the AutoText entry.

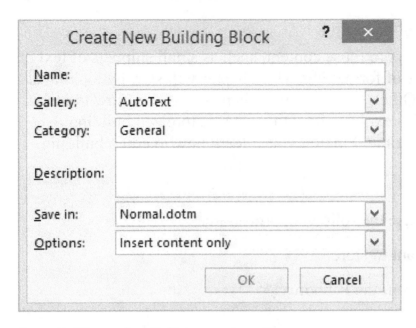

Step 7: Keep the Gallery as AutoText.

Step 8: Select a Category from the drop-down list. You can also create a new category to help organize your AutoText entries.

Step 9: Enter a Description, if desired, to explain the purpose of your AutoText entry.

Step 10: Select the template where you would like to save the AutoText entry from the Save in drop down list. Remember that Normal is the template used when you create a new blank document.

Step 11: Select an item from the Options drop down list. In most cases, you'll use Insert Content Only.

Step 12: Select OK.

Inserting a Quick Part

To insert a Quick Part, use the following procedure.

Step 1: Place your cursor where you want the reusable content to appear.

Step 2: Select the Insert tab from the Ribbon.

Step 3: Select Quick Parts.

Step 4: Select the Quick Part that you want to use from the Gallery. Or select AutoText and select the AutoText entry that you want to use from the AutoText Gallery.

The selected item is inserted at your cursor position.

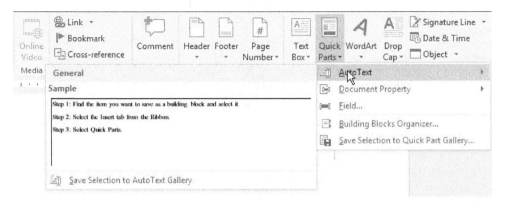

Creating Customized Building Blocks

To create a building block, use the following procedure.

Step 1: Find the item you want to save as a building block and select it.

Step 2: Select the Insert tab from the Ribbon.

Step 3: Select Quick Parts.

Step 4: Select Save Selection to Quick Part Gallery from the drop-down list.

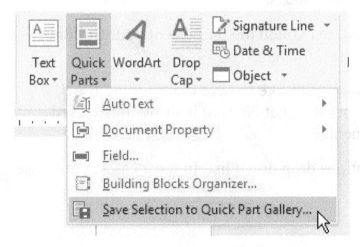

Step 5: In the Create New Building Block dialog box, enter a Name for the Building Block.

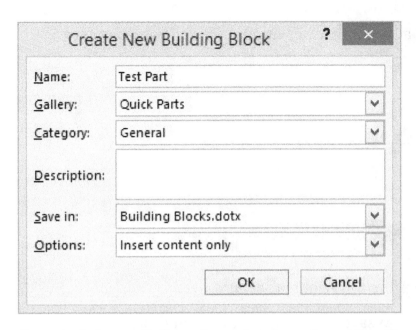

Step 6: Select a Gallery from the drop-down list. For example, if your building block is a footer, you can place it in the Footers gallery. However, in most cases, the default setting of Quick Parts is fine.

Step 7: Select a Category from the drop-down list. You can also create a new category to help organize your Building Blocks.

Step 8: Enter a Description, if desired, to explain the purpose of your Building Block.

Step 9: Select the template where you would like to save the Building Block from the Save in drop down list. Remember that Normal is the template used when you create a new blank document.

Step 10: Select an item from the Options drop down list. In most cases, you'll use Insert Content Only.

Step 11: Select OK.

Editing a Building Block

To review the Building Block Organizer, use the following procedure.

Step 1: Select the Insert tab from the Ribbon.

Step 2: Select Quick Parts.

Step 3: Select Building Block Organizer from the drop-down list.

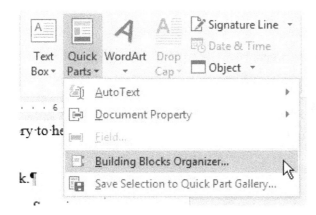

The Building Blocks Organizer lists all the Building Blocks that are associated with the template you are using for the current document. There are many built-in Building Blocks associated with themes and other items.

You can click on the headers at the top of the list to sort the items by Name, Gallery, Category, or Template. When you select a Building Block the right side of the window displays a preview of the Building Block.

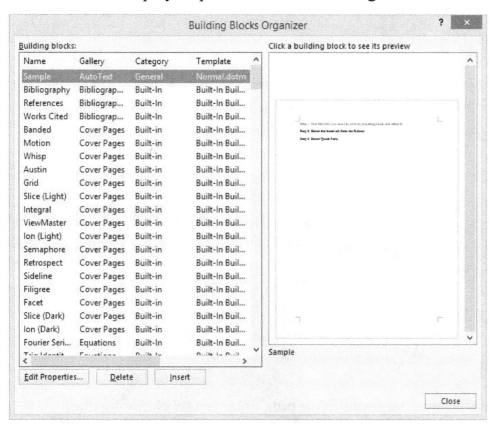

To delete a Building Block, highlight it in the list and select Delete.

To insert a Building Block, highlight it in the list and select Insert.

To edit the properties of a Building Block, use the following procedure.

Step 1: In the Building Blocks Organizer window, highlight the Building Block you want to edit, and select Edit Properties.

Step 2: In the Modify Building Blocks dialog box, edit the Name, Gallery, Category, Description, Template location, or Options.

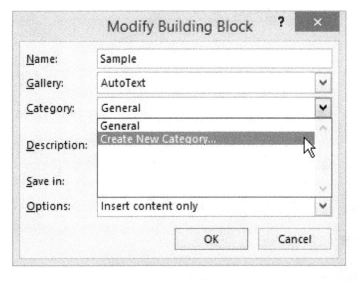

Step 3: To create a new Category, select the Category drop down list. Select Create New Category.

Step 4: In the Create New Category dialog box, enter a new Name for the Category. Select OK.

Step 5: In the Modify Building Block dialog box, select OK.

Step 6: Word displays a warning message to make sure that you want to redefine the Building Block. Select Yes to continue.

To correct a Building Block by saving over the original, use the following procedure.

Step 1: Insert the Quick Part you want to change into your document.

Step 2: Make the content and/or formatting changes that you want to make.

Step 3: Select the content to include in the corrected Quick Part.

Step 4: Select the Insert tab from the Ribbon.

Step 5: Select Quick Parts.

Step 6: Select Save Selection to Quick Part Gallery from the drop-down list.

Step 7: In the Create New Building Block dialog box, enter a Name for the Building Block. Make sure that the name is the same as the item you are correcting.

Step 8: Select the Gallery, Category, Description, Save in location, and Options as when you created the Building Block originally.

Step 9: Select OK.

Step 10: Word displays a warning message to make sure that you want to overwrite the Building Block entry. Select Yes to continue.

Chapter 23 – Working with Templates

Templates can be a huge time saver for occasions when you need to make many documents with the same types of formatting. In this module, you'll learn about using templates in Word 2016. Then you'll learn how to modify an existing template. You'll also learn how to create a new template. This module explains how to apply a template to an existing document so that you can quickly reformat a document. Finally, you'll learn how to manage your templates.

About Templates

Templates provide a consistent, familiar look to your documents. It is a convenient, time-saving way to create documents rather than modifying another document or starting from scratch each time. You may already be familiar with using the templates provided by Word.

Did you know that these templates could be modified so that you can create multiple documents from the modified template? You may want to start with a design from an Office.com template, but customize it for your purposes. Then you can create multiple documents from the modified template.

You can also create your own template. This allows you to completely customize the blueprint for the documents that will be based on your template. You can customize everything about the template, creating placeholders for styles, spacing and other design elements instead of adding content to the document.

Finally, you can even reformat a document that has already been created by applying a new template to the document. In this way, you can use different branding for different scenarios for the same document.

Here are some important tips to remember to maximize the benefit of using templates.

- Use Global settings instead of local ones. Global settings are the settings that affect the entire document (or most of it), such as themes and styles. Local settings are those that you have applied to a single object or paragraph. Just remember that you can quickly reformat a document by changing a style if styles were used consistently throughout the document. If the style is in the template from the beginning, it makes the formatting (or reformatting) that much easier.

- Similarly, use alignment and indents for paragraph spacing instead of using tabs or spaces. Tabs and spaces can cause problems when you replace placeholder text in a template with other content.

- Use tables for positioning items like graphics. Again, your spacing can be changed when you replace placeholder content with real content. Use a table with no borders instead and the spacing will stay the same for every document based on that template.

Modifying an Existing Template

To modify an existing template, use the following procedure.

Step 1: Open the file that you want to modify. The templates are stored in the following location:

C:\Users\user name\App Data\Roaming\Microsoft\Templates

Step 2: If there are not any files listed, select the Files drop down list and select Document Templates.

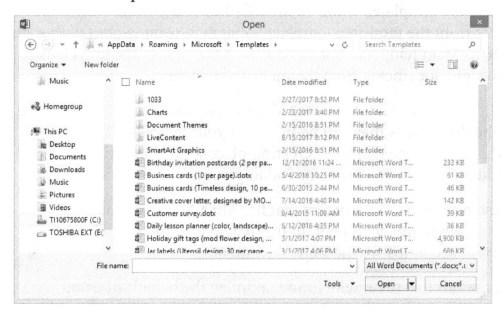

Step 3: Highlight the file you want to modify and select Open.

Step 4: Make the changes you want to have applied to future documents based on this template, including styles, page layouts, placeholder content, etc.

Step 5: Save the file.

Creating a New Template

To create a template, use the following procedure.

Step 1: Create a new, blank document.

Step 2: Make the changes you want to have applied to future documents based on this template, including styles, page layouts, placeholder content, etc. You can control any settings for the template to create consistency in future documents based on this template.

Step 3: Select the File tab from the Ribbon to open the Backstage View.

Step 4: Select Save As.

Step 5: Navigate to the following location in the Save As dialog box:

C:\Users\user name\App Data\Roaming\Microsoft\Templates

Step 6: Select Word Template or Macro-Enabled Word template (dotx or dotm) from the Save as Type drop down list.

Step 7: Select Save.

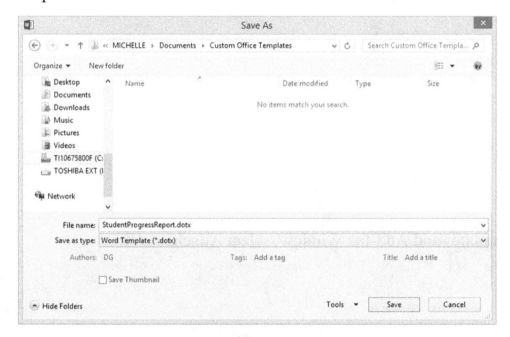

Applying a Template to an Existing Document

To apply a template to an existing document, use the following procedure.

Step 1: Open the document you want to reformat.

Step 2: Open the Options dialog box by selecting Options from the Backstage View.

Step 3: Select the Add Ins tab.

Step 4: Select the Manage drop down list. Select Templates from the list of options.

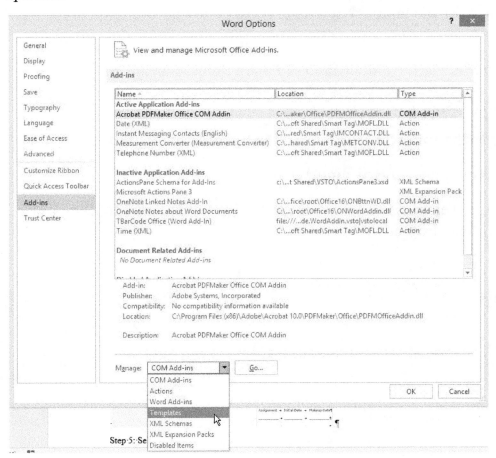

Step 5: Select Go.

Step 6: In the Templates and Add-Ins Window, select Attach.

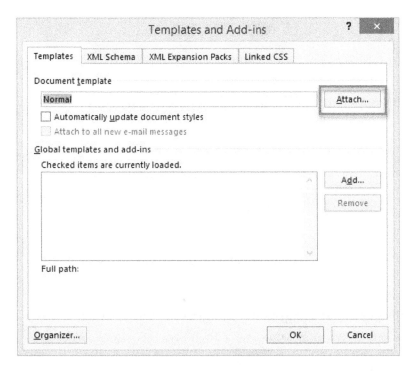

Step 7: In the Attach Template dialog box, navigate to the location of the template you want to apply. Highlight it and select Open.

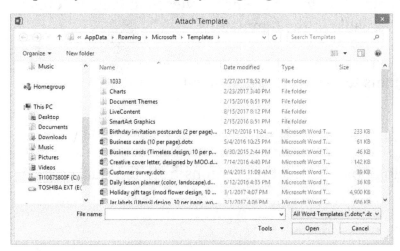

Step 8: In the Templates and Add-ins window, check the Automatically update document styles box to reformat the document using the templates styles.

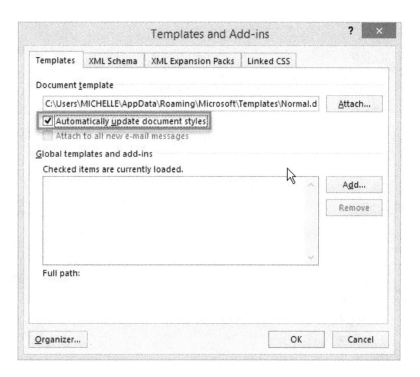

Step 9: Select OK.

The document is reformatted according to the themes, styles, and page layouts applied in the document and available in the template.

Managing Templates

To load additional template items to the current document, use the following procedure.

Step 1: Open the document you want to reformat.

Step 2: Open the Options dialog box by selecting Options from the Backstage View.

Step 3: Select the Add Ins tab.

Step 4: Select Templates from the Manage drop down list.

Step 5: Select Go.

Step 6: In the Templates and Add-Ins Window, select Add.

Step 7: In the Add template dialog box, select the template that contains the items you want to load to the current document. Select OK.

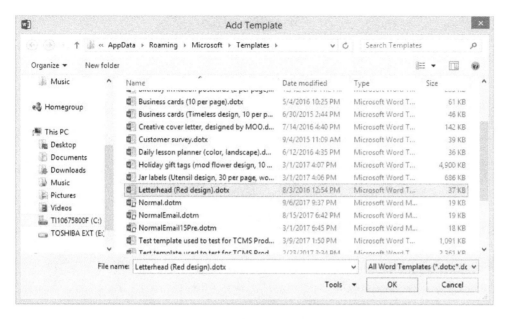

Step 8: The item is listed in the Templates and Add-ins window. Select OK to continue working in your document using the newly loaded items.

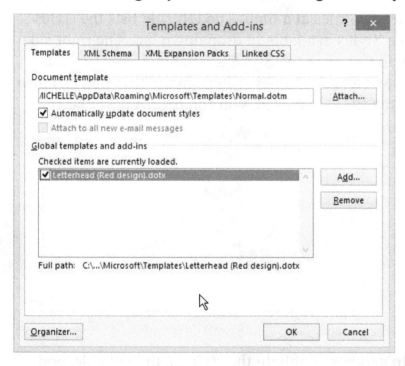

To use the Organizer, use the following procedure.

Step 1: In the Templates and Add-ins window, select Organizer.

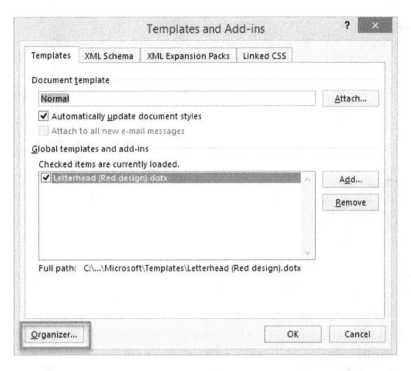

The Organizer window includes two files at a time. The left side lists the styles available in the file listed in the Styles available in drop down list. The right side lists the styles available in the file listed in the Styles available drop-down list on the right side. On either side, you can select a new file from the list. You can also close a selected file. When all the files are closed, you can open a new one.

To copy styles from one file to another, highlight the style on the left side, and select Copy. You can also highlight a style and delete it or rename it. Select Close when you have finished working in the Organizer window.

Chapter 24 – Working with Sections and Linked Content

In this chapter, you'll learn some powerful uses for sections. First, we'll take a general look at sections and learn how to enter a section break. Next, we'll cover how to customize page numbers in a document using sections. Then we'll look at using multiple page formats in a document. This module also explains how to use different headers and footers in a document. Finally, we'll look at how to link and unlink text boxes.

Using Sections

To insert a section into a document, use the following procedure.

Step 1: Place the cursor in the location where you want to split the document. The new section will begin where the cursor is located.

Step 2: Select the Layout tab from the Ribbon.

Step 3: Select the type of Section Break from the drop-down list.

Next Page – select this option to start the section on the next page. You'll need this one if you want to use the section to create different page layouts within the document.

Continuous – select this option to start the section immediately. You might use this one if you want to include different column layouts within the same page.

Even Page – select this option if you are using a two-page layout and you want the next section to start on an even page. A blank page will be inserted if necessary.

Odd Page – select this option if you are using a two-page layout and you want the next section to start on an add page. A blank page will be inserted if necessary.

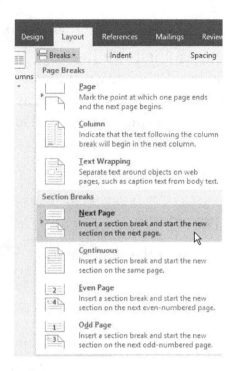

Customizing Page Numbers in Sections

To create custom page numbers, use the following procedure.

Step 1: Double-click in the footer area of the first section to open the Header & Footer Tools Design tab on the Ribbon.

Step 2: If the Link to Previous option is active (in the Navigation group), select it to turn it off. Customized page numbers do not work if the sections are linked.

Step 3: Enter the page number in the desired location by selecting Page Number and select the desired option from the drop-down list.

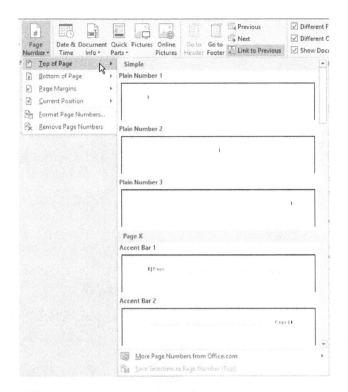

Step 4: Select Format Page Numbers from the Page Number drop down list to open the Page Number Format dialog box.

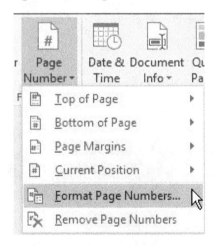

Step 5: Select the Number Format from the drop-down list.

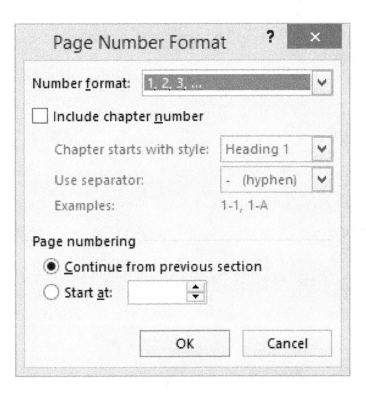

Step 6: Select Start at and enter the starting page number for this section.

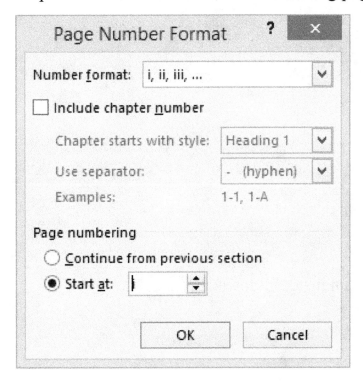

Step 7: Select OK.

Step 8: Make sure there is a section break at the end of the current section. Move to the next section's footer. If the Link to Previous option is active (in the

Navigation group), select it to turn it off. You may need to unlink each section separately.

Step 9: Select Format Page Numbers from the Page Number drop down list to open the Page Number Format dialog box for this section.

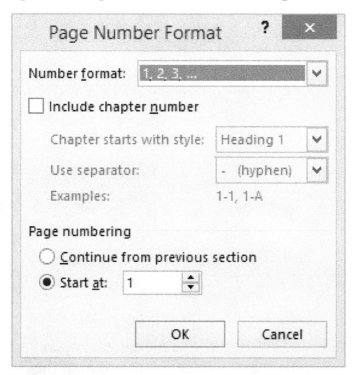

Step 10: Choose the Number format and the Page numbering start location for this and select OK to apply the formatting to this section's page numbering.

Using Multiple Page Formats in a Document

To add a landscape section to a document that is portrait oriented, use the following procedure.

Step 1: Create a section break at the end of the document.

Step 2: Select the Layout tab on the Ribbon.

Step 3: Select the Breaks expanded menu option.

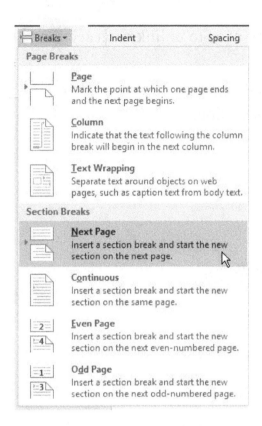

Step 4: Select Next Page.

Step 5: Making sure that the cursor is located AFTER the section break, open the Layout dialog box by selecting the small square in the Page Setup group of the Layout tab on the Ribbon.

Step 6: On the Margins tab, select Landscape as the orientation.

Step 7: In the Apply To list at the bottom, make sure that This Section is selected from the drop-down list.

Step 8: Select OK.

The new section has a different page orientation.

Using Different Headers and Footers in a Document

To use different headers and footers using sections, use the following procedure.

Step 1: Double-click in the header area of the first section to open the Header & Footer Tools Design tab on the Ribbon.

Step 2: In this example, the Title Page should not have headers or footers, so we'll check the Different First Page box. Select Go to Footer and check the Different First Page box for it.

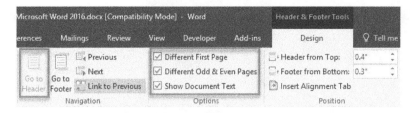

Step 3: If the Link to Previous option is active (in the Navigation group), select it to turn it off. It is highlighted if it is active. The Link to Previous option makes the active header or footer the same as the previous section's header or footer. You'll need to unlink headers and footers separately.

Step 4: Make sure there is a section break at the end of the current section. Move to the next section's header or footer. If the Link to Previous option is active (in the Navigation group), select it to turn it off. You may need to unlink each section separately.

Step 5: Enter the header and/or footer information that is different from the previous section.

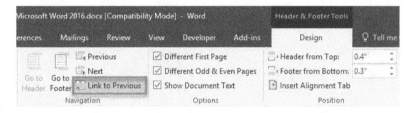

Linking and Breaking Links for Text Boxes

Use the following procedure to link text boxes.

Step 1: Scroll to the text box on the page.

Step 2: Click the text box to select it and open the Text Box Tools.

Step 3: Select the Format tab on the Text Box Tools Ribbon.

Step 4: Select Create Link.

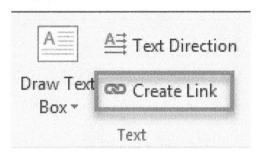

Notice how the cursor changes to a pitcher. This indicates that you are creating a text box link, and there is text to flow to an empty text box.

Step 5: Click on the empty text box where you want the text to flow. Notice how the cursor changes to a pouring pitcher when you mouse over an empty text box.

Step 6: The text boxes are now linked. Extra text from the first text box flows into the second text box.

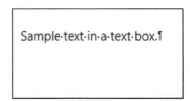

To break a link, return to the first text box. When you select the text box, the Break Link option becomes available.

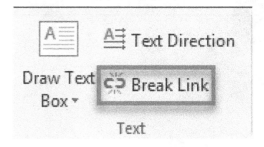

Word 2016 has some great features to help you work with your documents. If you need to go back to an earlier version, you can use the auto save feature to help you restore an earlier version. This module explains how to configure your auto save settings so that previous versions of your documents will be available. It also explains how to review, compare, and restore previous versions. You'll also learn how to work with tracked comments and changes from multiple authors. First, you'll learn how to combine the changes and comments into one document. Then, you can use that compilation to review all the comments at one time.

Merging Different Versions of a Document

To configure the auto-save settings, use the following procedure.

Step 1: Select the File tab from the Ribbon to display the Backstage view.

Step 2: Select Options.

Step 3: In the Word Options dialog box, select the Save tab.

Step 4: Check the Save AutoRecover information every __ minutes box to enable the auto save feature.

Step 5: Enter a number of minutes in between auto saves in the box, or you can use the up and down arrows to adjust the number of minutes.

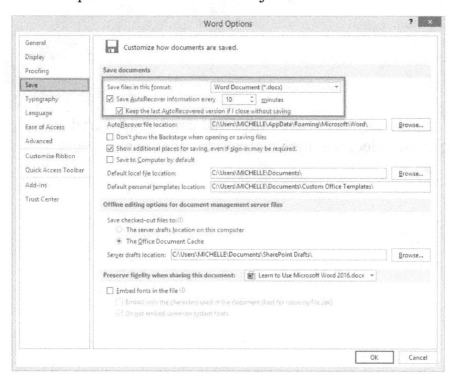

Step 6: Select OK.

To view the Versions on the Backstage view, select the File tab on the Ribbon. Make sure that the Info tab on the Backstage view is selected.

You can click on a version to open it as a separate file. A message appears at the top of the file that indicates it is an AutoSaved version.

Select Compare to open a new file with changes marked between the version you selected and the original file you have open.

Select Restore to return the selected version to the original file you have open.

To recover an unsaved document, use the following procedure.

Step 1: Select the File tab from the Ribbon to display the Backstage view.

Step 2: Select Manage Document on the Info menu, and then select Recover Unsaved Documents.

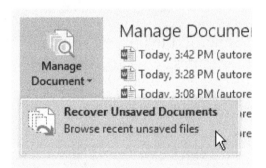

Step 3: In the Open dialog box, any auto-saved files that can be recovered are shown in the default location. Select the one you want to recover and select Open.

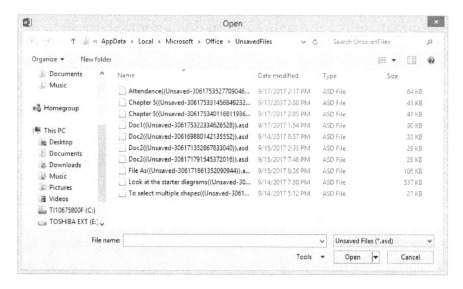

Tracking Comments in a Combined Document

To merge comments and changes from several documents into one document, use the following procedure.

Step 1: The file where you want to combine your changes should be open.

Step 2: Select the Review tab from the Ribbon.

Step 3: Select the Compare expanded menu and then select Combine from the drop-down list.

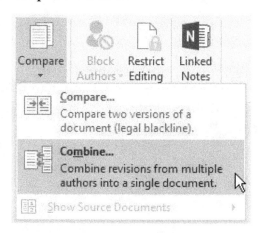

Step 4: In the Original document area, select the name of the document where you want to combine the changes from multiple sources. If it isn't open, select the folder to open the file.

Step 5: Make sure that any changes in this document are marked with a name or initials by entering the desired Label.

Step 6: Under Revised document, select the name of the document where the changes are from the drop-down list (if the file is open). Otherwise, select the folder to open the file.

Step 7: Make sure that any changes in this document are marked with a name or initials by entering the desired Label.

Step 8: If you need to switch the documents (you have the document with revisions as the original), select the double arrow icon.

Step 9: Select More to see all the Comparison Settings.

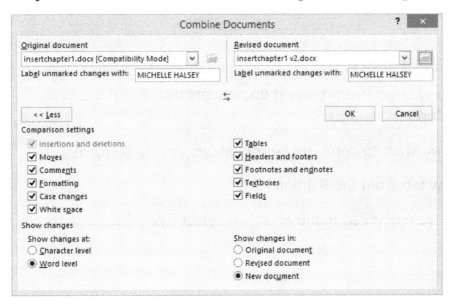

Step 10: Check the boxes to indicate which items to include in the comparison.

Step 11: Indicate whether you want to show changes at the Character or Word level. For example, if the word cat is changes to cats, Word shows the entire word changed instead of just the letter s by default.

Step 12: Indicate whether to show the changes in the original document, the revised document, or a new document.

Step 13: Select OK.

Step 14: For multiple authors, repeat steps 1-13 until you have merged all the changes into a single document.

Reviewing Comments in a Combined Document

To review the comments in the combined documents, use the following procedure.

Step 1: In the document where the comments have been combined, select the Review tab from the Ribbon.

Step 2: Select Reviewing Pane. Select Reviewing Pane Vertical to see the comments on the left side of the Word window. Select Reviewing Pane Horizontal to see the comments on the bottom of the Word window.

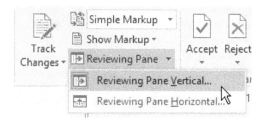

Step 3: The Revisions pane color codes the comments, with the name or initials of the author on the right side of the comment/change heading. Each change or comment is marked with the type of change requested, such as Deleted, Inserted, or Comment. You can make edits right in the Revisions pane.

Step 4: You can also see the comments in bubbles on the right side of the Word window. Remember that you can respond to comments right in the Comments window.

218

Chapter 26 – Using Cross References

In this chapter, you'll learn how to use cross references to guide your reader to other information in your document. First, we'll learn about the different types of cross references. Then, you'll learn how to insert a bookmark to use as a cross reference. This module explains how to insert a cross reference to a bookmark or to heading text. You'll also learn how to update a cross reference. Finally, we'll look at some advanced tools to use in formatting your cross references.

Types of Cross References

Discuss the different types of cross references.

- Numbered items – references a selected paragraph number.
- Headings – references a selected paragraph formatted with a heading style.
- Bookmarks – references a bookmark location you have inserted into the document.
- Footnotes – references a selected footnote.
- Endnotes – references a selected endnote.
- Equations – references a selected equation.
- Figures – references a selected figure.
- Tables – references a selected table.

Inserting a Bookmark

To insert a bookmark, use the following procedure.

Step 1: Place your cursor in the location where you want to insert the bookmark.

Step 2: Select the Insert tab from the Ribbon.

Step 3: Select Bookmark.

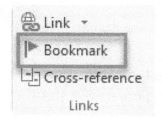

Step 4: In the Bookmark dialog box, enter a Bookmark Name for your location. This name will help you find this location later.

Step 5: Select Add.

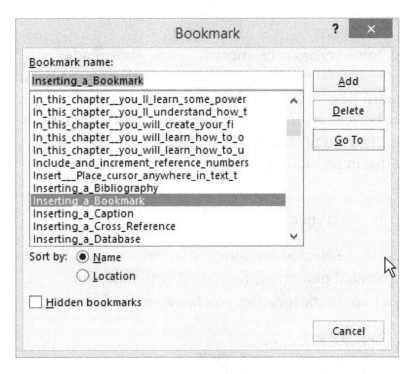

Inserting a Cross Reference

To insert a cross reference, use the following procedure.

Step 1: Place your cursor in the location where you want to insert the cross reference.

Step 2: Select the Insert tab from the Ribbon.

Step 3: Select Cross-reference.

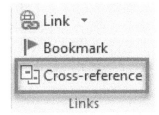

Step 4: In the Cross-reference dialog box, select the Reference type from the drop-down list. In this example, we'll choose the Bookmark we created in the previous lesson.

Step 5: Select the bookmark you want to use from the for which bookmark list.

Step 6: Select the type of information you want to reference from the Insert reference to drop down list. In this case, we want to use the page number.

Step 7: Select Insert.

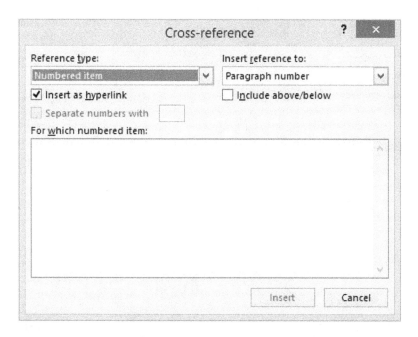

To insert a cross reference that includes heading text, use the following procedure.

Step 1: Place your cursor in the location where you want to insert the cross reference.

Step 2: Select the Insert tab from the Ribbon.

Step 3: Select Cross-reference.

Step 4: In the Cross-reference dialog box, select the Reference type from the drop-down list. In this example, we'll choose a heading.

Step 5: Select the heading you want to use from the for which Heading list.

Step 6: Select the type of information you want to reference from the Insert reference to drop down list. In this case, we want to use the heading text.

Step 7: Select Insert.

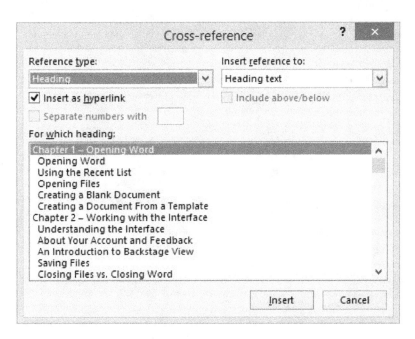

To update a single cross reference, use the following procedure.

Step 1: Right-click the cross-reference text. The cross-reference text will be highlighted in gray when you select it or right-click on it.

Step 2: Select Update Field from the context menu.

To update all fields in a document at once, use the following procedure.

Step 1: Press Ctrl + A to select all text in the document.

Step 2: Press F9.

Step 3: If the document has a Table of Contents or other special types of fields, you may get a confirmation message to clarify how you want to update the fields.

Note that using this method does not update fields in the header or footer. You will need to select cross references placed there separately.

Formatting Cross References Using Fields

Right-click a field and select Edit Field from the context menu to open the Fields dialog.

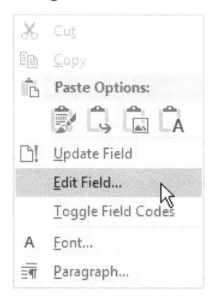

Review the Field dialog box.

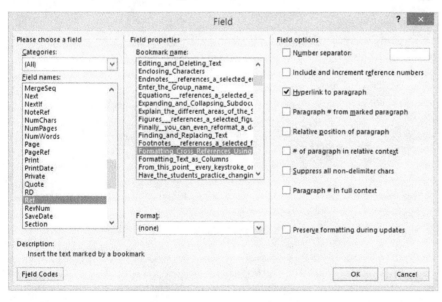

The current Field Name is selected with the Field properties, such as the location it references. The Field options include additional ways you can control the field, including:

- Number separator

- Include and increment reference numbers

- Hyperlink to paragraph

- Paragraph # from marked paragraph

- Relative position of paragraph

- # of paragraph in relative context

- Suppress all non-delimiter chars

- Paragraph # in full context

- Preserve formatting during updates

The Format drop down list allows you to control the case of the reference.

The Field Options are different, depending on what type of field you are editing.

Select Field Codes to open the Advanced field properties area where you can edit the actual coding.

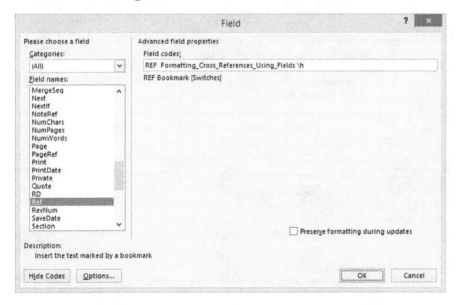

Select Options to see the switches.

The General Switches tab includes the same case formatting options you've seen previously. When you select an option, you can see the description at the bottom. Select Add to Field to include the switch with the field code.

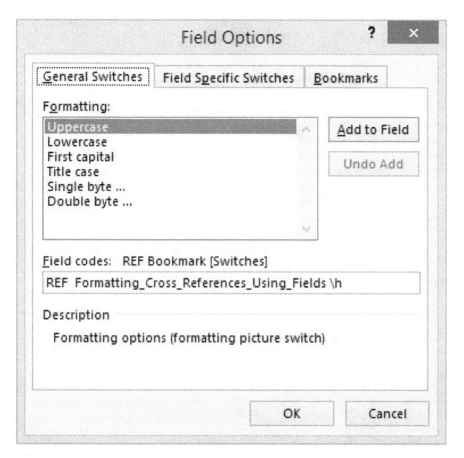

The Field Specific Switches tab includes additional options, based on what type of field you have selected. When you select an option, you can see the description at the bottom. Select Add to Field to include the switch with the field code.

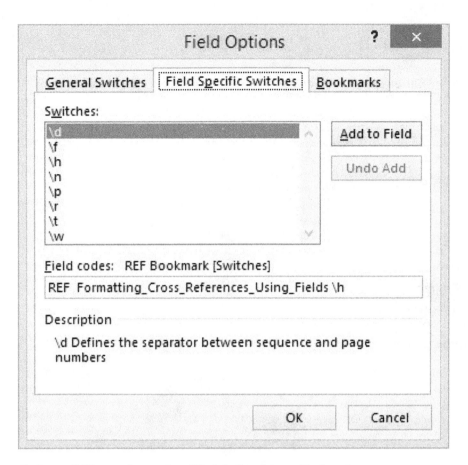

Select OK to close the Field Options dialog box.

Chapter 27 – Creating Mail Merges and Labels

This chapter explains how to use mail merges to create customized letters, emails and labels. You'll learn how to create a mail merge with an external data source. Then you'll learn how to create a custom merge by entering a new list of data for barcodes. This module also explains how to create return address labels using the Labels option. Finally, you'll learn about using Avery Label Templates.

Creating a Mail Merge

To create a mail merge, use the following procedure.

Step 1: Open the document that contains the letter you want to personalize in a mail merge.

Step 2: Select the Mailings tab from the Ribbon.

Step 3: Select Start Mail Merge. Select Step by Step Mail Merge Wizard from the drop-down list.

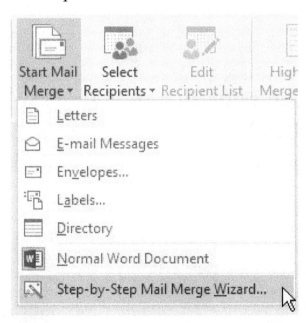

Step 4: The Mail Merge pane opens on the right side of the screen.

Step 5: Indicate the Document Type you want to use. In this example, we'll keep Letter selected. Note that you can create emails, envelopes, letters, or a directory.

Step 6: Select Next at the bottom of the Mail Merge pane.

Step 7: Indicate which Starting Document you want to use. In this example, we'll use the Current Document.

Step 8: Select Next.

Step 9: Select the Recipients. In this example, we'll use an existing Excel file.

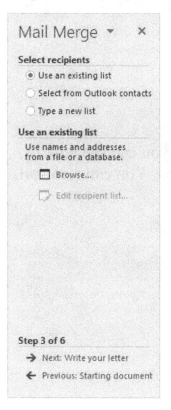

Step 10: Select Browse to open the file.

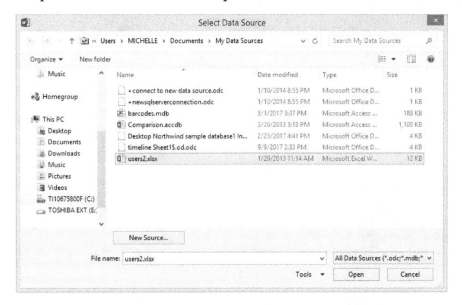

Step 11: Select the location of the file and select Open.

Step 12: Select Table and then select Ok.

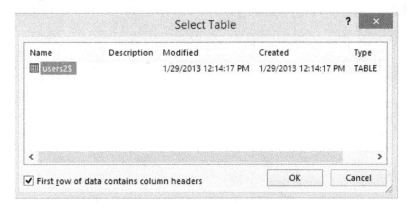

Step 13: The Mail Merge Recipients window opens, so that you can see how Word imported the columns. You can select or unselect items. You can also sort, filter, find and validate information. Select OK.

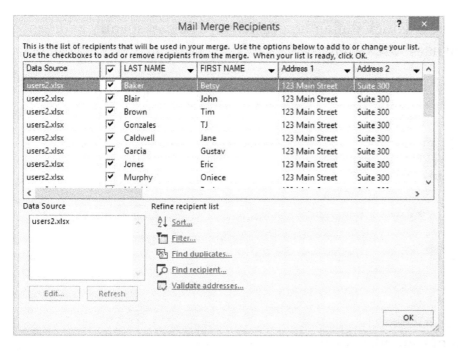

Step 14: Select Next.

Step 15: The Write Your Letter step on the wizard allows you to add the personalized details to your letter. Place your cursor in the location on the letter where you want the address to appear. Select Address Block to add the contact's address block to the letter.

Step 16: The Insert Address Block dialog box appears. You can specify address elements and preview what those details will look like. Select OK when you have finished.

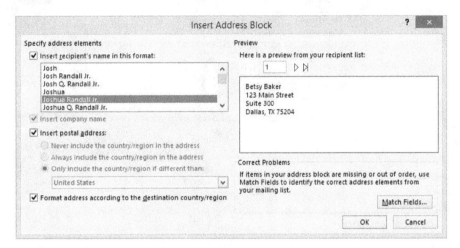

Step 17: Word enters a placeholder for the Address Block.

Step 18: Now let's add a personalized greeting. Move your cursor to the location where you want the greeting. Select Greeting Line from the Mail Merge pane.

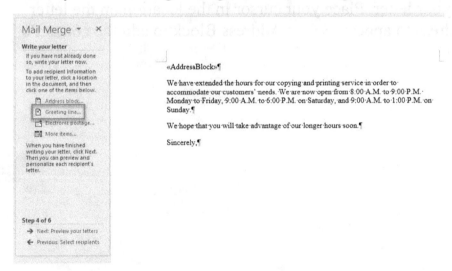

Step 19: The Insert Greeting Line dialog box appears. You can specify the formats and preview what those details will look like. Select OK when you have finished.

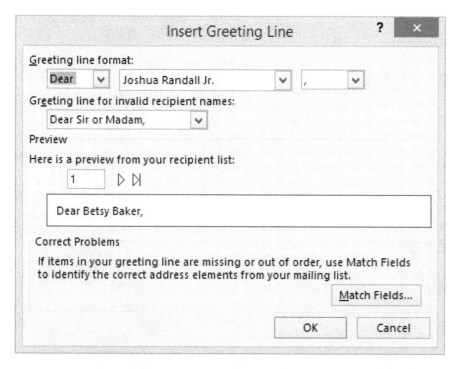

Step 20: Now select Next to preview the results. Notice the Tools in the Mailings ribbon to help you navigate through the list. You also can navigate using the Mail Merge pane.

Step 21: Select Next.

Step 22: Select Print to print your letters.

Step 23: The Merge to Printer dialog box allows you to select All, the Current record, or a page range. Select OK.

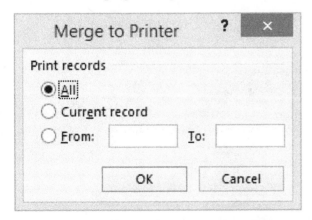

Step 24: The Print dialog allows you to control the printing options. Select OK when you are ready to print.

Note: The Print dialog pulls from the printer driver installed on your computer. This screen will vary based on this driver.

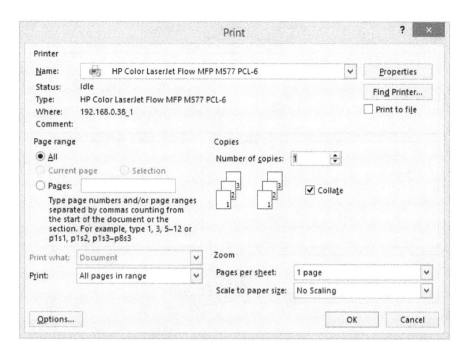

Creating Return Address Labels

To store their return addresses and introduce the Envelopes option, use the following procedure.

Step 1: Select the Mailings tab from the Ribbon.

Step 2: Select Envelopes.

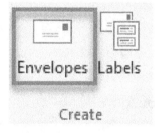

Step 3: In the Return Address area, enter your return address.

Step 4: Select Add to Document.

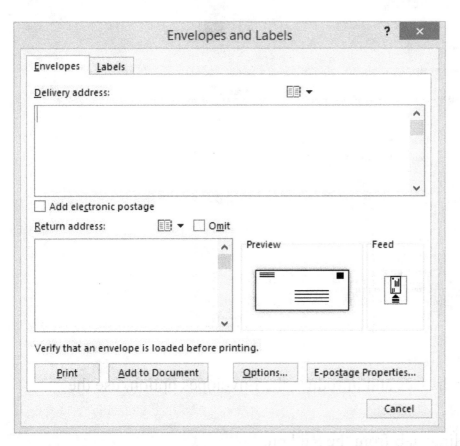

When Word prompts you to save the default return address, select Yes.

To create a sheet of return address labels, use the following procedure.

Step 1: Select the Mailings tab from the Ribbon.

Step 2: Select Labels.

Step 3: Check the Use Return Address box to display the return address you have previously saved (or if you want to enter a new address and save it as the default return address). You can also just enter the address you want to use in the Address box. You can also choose the Address Book icon to select an address from Outlook.

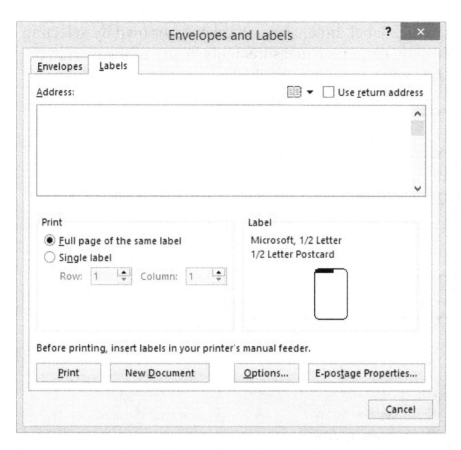

Step 4: To create a sheet of return address labels, make sure that Full page of the same label is selected.

Step 5: Select Options to choose the type of label you are using. In the Label Options dialog box, select the Printer Information, the Label Vendor, and the Product Number.

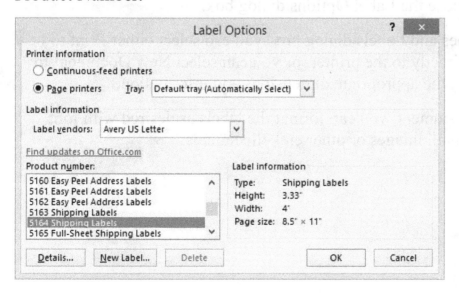

Step 6: You can also customize label dimensions or create your own by selecting Details or New Label. You can enter new measurements in any of the margin or dimension fields. Select OK when you have finished.

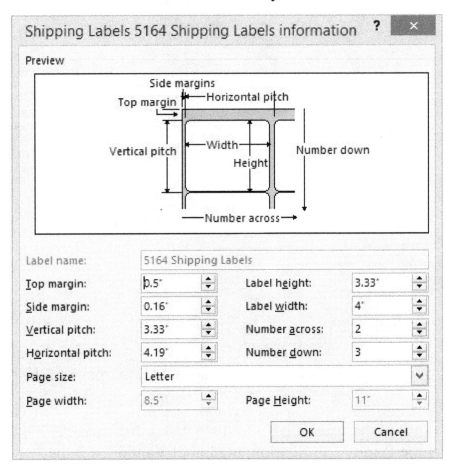

Step 7: Select OK to close the Label Options dialog box.

Step 8: In the Envelopes and Labels dialog box, you can select either Print to send the plain labels directly to the printer, or you can select New Document to create a document with the appropriate dimensions and information.

If you select a New Document, you can format the labels as desired with font changes or even add small images or other embellishments.

Anywhere,·USA¶
123·Main·Street¶
City,·State·Zip¤

Anywhere,·USA¶
123·Main·Street¶
City,·State·Zip¤

Anywhere,·USA¶
123·Main·Street¶
City,·State·Zip¤

Anywhere,·USA¶
123·Main·Street¶
City,·State·Zip¤

Master documents allow you to keep track of a few related documents and combine them in a single place to control page numbering, printing, and other activities. In this module, you'll learn how to create a master document and create subdocuments. You'll also learn how to insert a subdocument. Then we'll cover how to work with subdocuments, including expanding and collapsing the subdocuments in the master document, unlinking a subdocument, and merging and splitting subdocuments. Finally, we'll look at how to lock a master document so that changes are not saved in the subdocuments accidentally.

Creating a Master Document

To create a master document, use the following procedure.

Step 1: Start with a blank document.

Step 2: Select the View tab from the Ribbon.

Step 3: Select Outline.

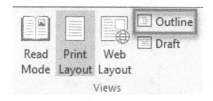

Now you are ready to work with your master document.

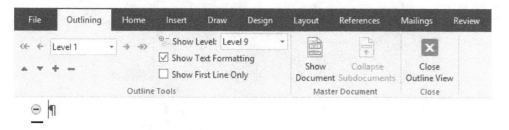

Creating Subdocuments

To create subdocuments in a master document, use the following procedure.

Step 1: On the Outlining tab of the Ribbon, select Show Document to show the tools for working with subdocuments.

Step 2: Enter some simple headings in your outline view of the sample document, such as:

- Chapter 1

  o Heading 1

  o Heading 2

- Chapter 2

  o Heading 1

  o Heading 2

Step 3: Highlight all the text.

Step 4: Select Create from the Outlining tab on the Ribbon.

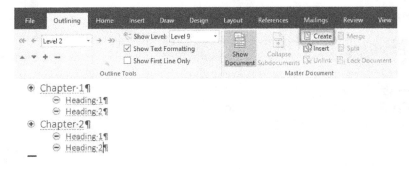

Step 5: Each "Chapter" becomes its own subdocument. Each subdocument is surrounded by a box and separated by a section break in the master document.

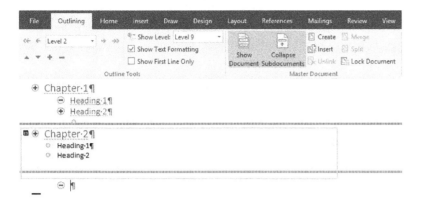

When you save the master document, each subdocument is saved as its own file. The files are named with the text of the heading used at level one for each subdocument.

Note that when you make changes to the text from the master document, those changes are also saved in the affected subdocument file.

Inserting a Subdocument

To insert a subdocument, use the following procedure.

Step 1: On the Outlining tab of the Ribbon, select Show Document to show the tools for working with subdocuments.

Step 2: Select Insert.

Step 3: In the Insert Subdocument dialog box, navigate to the location of the file you want to use as a subdocument. Highlight it and select Open.

Step 4: If the master document and the subdocument use different templates, you will get a warning message. Select OK. Or if they should use the same template, then you will need to go back and use the appropriate template when creating your master document.

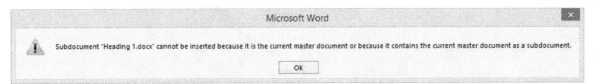

Step 5: If the master document and the subdocument use different templates, and they include styles with the same name, but different formatting, you will get an additional warning message. Select Yes to All to rename the styles, or No to All to keep the same names, which will help with reformatting if you are applying a new template.

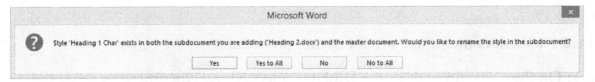

The subdocument is inserted. Notice that there is a section break automatically inserted at the end of the subdocument.

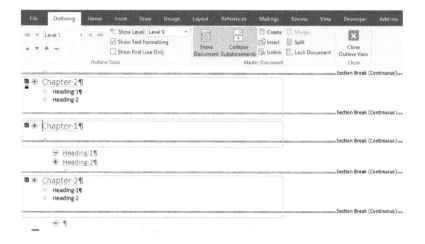

Expanding and Collapsing Subdocuments

To collapse or expand the subdocuments in the master document, use the following procedure.

Step 1: With the text of the subdocuments showing, select Collapse Subdocuments from the Outlining tab on the Ribbon.

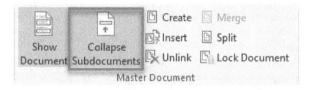

The collapsed view shows the document reference instead of the contents.

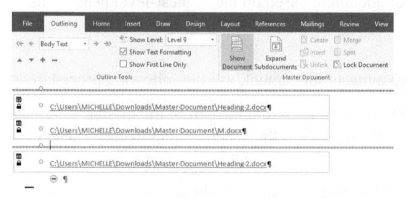

Step 2: With the text of the subdocuments collapsed, select Expand Subdocuments from the Outlining tab on the Ribbon to see the text again.

Merging and Splitting Subdocuments

To merge subdocuments, use the following procedure.

Step 1: Select the subdocuments in your master document that you want to merge. Notice the small square in the top of each subdocument box. If you click there, it will select the entire subdocument. Hold down the SHIFT or CTRL key to select multiple subdocuments.

Step 2: Select Merge from the Outlining tab on the Ribbon.

Step 3: Notice that the subdocument icon has been removed from the second subdocument. When you save the master document, the affected subdocuments are also saved.

To split subdocuments, use the following procedure.

Step 1: Select the text in your master document that you want to split from its subdocument into a different subdocument.

Step 2: Select Split from the Outlining tab on the Ribbon.

Step 3: Word creates a new subdocument based on the highest-level heading of the text you selected.

Unlinking a Subdocument

To unlink a subdocument from a master document, use the following procedure.

Step 1: Select the subdocument that you want to unlink from the master document. Use the little icon at the top left of the subdocument to easily select the whole subdocument.

Step 2: Select Unlink from the Outlining tab on the Ribbon.

Locking a Master Document

To lock a master document, use the following procedure.

Step 1: Select Lock Document from the Outlining tab on the Ribbon.

Step 2: Notice the lock icon shown with each subdocument icon.

Step 3: To unlock the subdocuments, select Lock Document again.

Macros allow you to automate frequently used tasks. You can use macros to speed up routine editing and formatting or combine multiple commands. You can even use a macro to make an option in a dialog box more accessible. This module focuses on learning how to record a macro and how to run a macro. We'll also cover how to apply macro security. Finally, we'll learn how to assign a macro you have recorded to a command button so that it is available from the Ribbon.

Recording a Macro

To record a macro, use the following procedure.

Step 1: Select the View tab from the Ribbon.

Step 2: Select Macros.

Step 3: Select Record Macro.

The Record Macro dialog box is displayed.

Step 4: Enter a Name for your macro. The name cannot contain spaces.

Step 5: Select the location where you would like to Store your macro from the drop-down list.

Step 6: If desired, enter a Description of what your macro accomplishes.

Step 7: Select OK.

From this point, every keystroke or command that you perform is recorded. Keep that in mind, so you don't accidentally record things that you don't want performed repetitively. You can type text, perform formatting or insert things like pictures or tables. Just about anything you can do in Word can be recorded in a macro.

Your cursor changes to an icon that looks like a cassette tape...a relic from the first days of macros in previous versions of Word.

Step 8: For this example, insert a table and apply formatting.

Step 9: When you have finished recording your actions, select the View tab from the Ribbon again. Select Macros. Select Stop Recording.

Note that you can also Pause Recording to correct something that you don't want as part of your macro, then Resume Recording when you are ready.

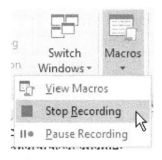

Running a Macro

To run a macro, use the following procedure.

Step 1: Select the View tab from the Ribbon.

Step 2: Select Macros.

Step 3: Select View Macros.

Step 4: In the Macros dialog box, select the Macro that you want to run. You can see the Description at the bottom to make sure it's the right one.

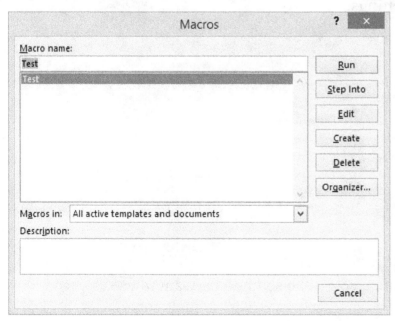

Step 5: Select Run.

Remember that a macro is a series of commands performed separately in the recording, though you get the result almost instantaneously. That means that when you select Undo, it only applies to the last command the macro performed.

Applying Macro Security

To change the macro security, use the following procedure.

Step 1: Select the File tab to open the Backstage view.

Step 2: Select Options.

Step 3: In the Word Options dialog box, select Trust Center.

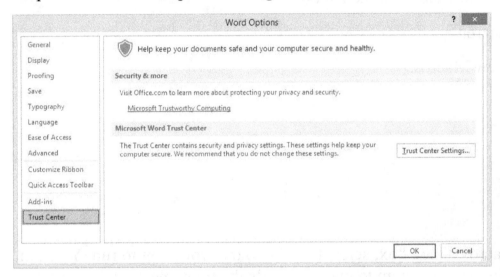

Step 4: Select Trust Center Settings.

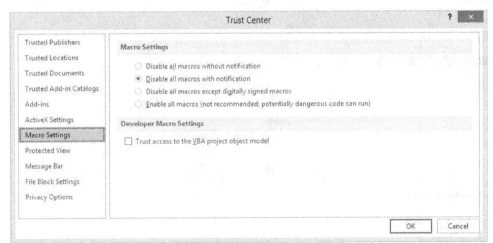

Step 5: In the Macro settings area, select the setting you want to use.

Step 6: Select OK.

To assign a new macro to a command key, use the following procedure.

Step 1: Select the View tab from the Ribbon.

Step 2: Select Macros.

Step 3: Select Record Macro.

The Record Macro dialog box is displayed.

Step 1: Enter a Name for your macro. The name cannot contain spaces.

Step 2: Enter a Description for your macro, if desired.

Step 3: Select Button.

Word opens the Word Options dialog box, open to the Quick Access Toolbar tab.

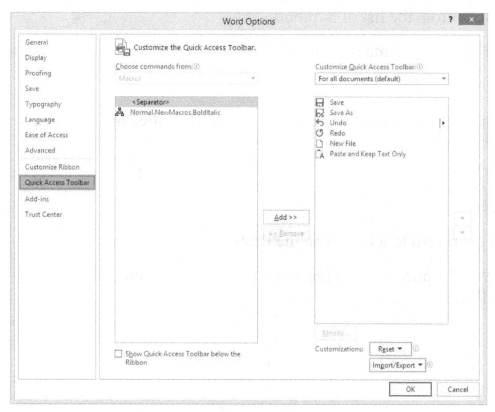

Your Macro is shown in the left list, you can add it to the Quick Access Toolbar by highlighting the macro and selecting Add. To change the icon or name, select Modify.

Step 1: Select an icon from the list of Symbols.

Step 2: Enter a new Display name if desired.

Step 3: Select OK.

Now you are recording your macro, as previously learned. Notice the icon you selected in the Quick Access Toolbar.

You can also add the macro to a button on the Ribbon.

Step 1: With the Word Options dialog box open (as in the procedure above), select Customize Ribbon.

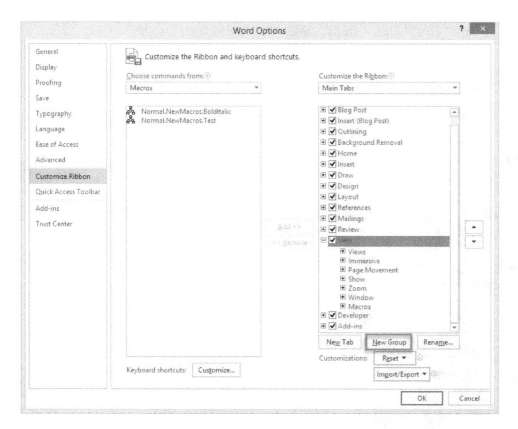

Step 2: Select Macros from the Choose Commands from drop down list.

Your Macros are shown in the left list. You must add a custom group where you will place the button for your macro.

Step 1: Select New Group.

Step 2: Select Rename.

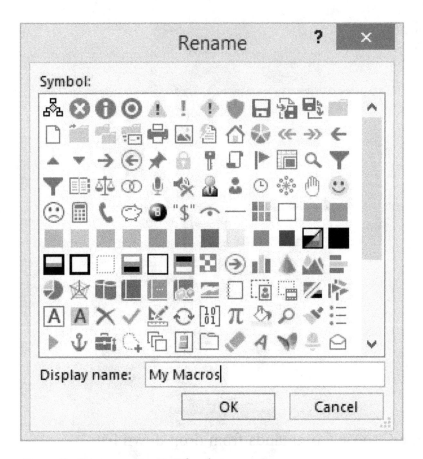

Step 3: Enter a new Display name.

Step 4: Select OK.

Now add the macro to the group. Select it in the list on the left and make sure you have your custom group selected on the right.

Step 1: Select Add.

Step 2: Select Rename.

Step 3: Select an icon for the macro from the list of Symbols.

Step 4: Select OK.

To assign a new macro to a keyboard shortcut, use the following procedure.

Step 1: Select the View tab from the Ribbon.

Step 2: Select Macros.

Step 3: Select Record Macro.

The Record Macro dialog box is displayed.

Step 4: Enter a Name for your macro. The name cannot contain spaces.

Step 5: Enter a Description for your macro, if desired.

Step 6: Select Keyboard.

The Customize Keyboard dialog box is displayed.

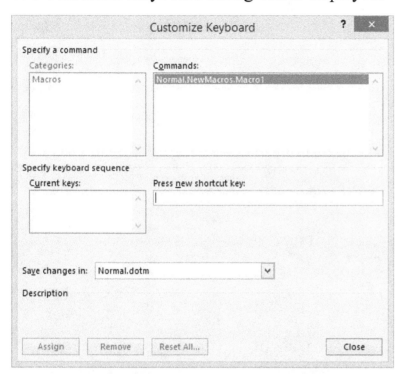

Step 7: In the Press new shortcut key field, press the keys you would like to use for running your macro. If that key is already assigned, it will show the command that key combination is currently used for. You can overwrite a previous association or choose a different key combination. Keys that are pressed at the same time will show a plus sign between them. Keys that are pressed in sequence will show a comma between them.

Step 8: Select Close.

Continue recording your macro.

To run the macro, you'll only need to press the shortcut key combination you selected.

Chapter 30 – Working with Forms

In this chapter, you'll learn about forms, where you can make it easy for users to enter specific data without changing the look or spacing of your document. In the first lesson, you'll learn about the Developer tab and creating a form from a template. Then, you'll learn about the form controls, which allow you to add different types of controlled content. This module also explains how to lock a form and add or remove fields. Finally, you'll learn how to insert data from a database onto a form.

Displaying the Developer Tab

The sample form has some areas where the user is asked to change or add information. In the next lesson, we'll learn how to change these to content controls.

To open the Developer tab, use the following procedure.

Step 1: Open the Options dialog box by selecting Options from the Backstage View.

Step 2: Select the Customize Ribbon tab.

Step 3: Check the box next to Developer.

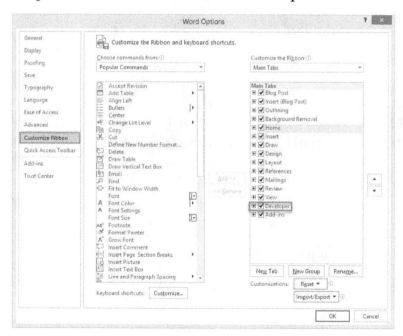

To review the Developer tab hover the mouse over the commands to see the screen tips.

Using Form Controls

To add a text control to a template, use the following procedure.

Step 1: On the Developer tab of the Ribbon, select Design Mode.

Step 2: Place your cursor in the document template where you want to text control to appear.

Step 3: Select Rich Text Content Control (to allow users to format their text) or the Plain Text Content Control. Word inserts the content control on the document template.

You can format the content control as needed.

Step 4: Make sure to turn off Design Mode when you have finished. Save your changes to the template.

Have the students practice changing items in the sample form to content controls, such as text boxes and check boxes. Turn off Design Mode and practice filling out the form.

Locking and Unlocking a Form

To group the contents of the form, use the following procedure.

Step 1: Select all the text on the form by pressing Ctrl + A.

Step 2: Select Group from the Developer tab on the Ribbon.

Step 3: Select Group.

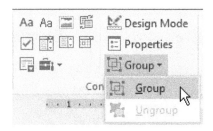

Only the Content Control areas can be changed now.

Step 4: Select Group again to remove the grouping.

Use the following procedure to review the Content Control properties.

Step 1: Select the text in a Content Control

Step 2: Select Properties.

Investigate the difference between checking each of the Locking option boxes.

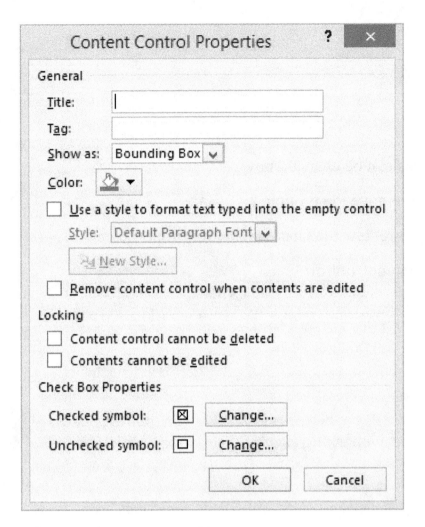

To add a field, use the following procedure.

Step 1: Select the Insert tab from the Ribbon.

Step 2: Select Quick Parts.

Step 3: Select Field.

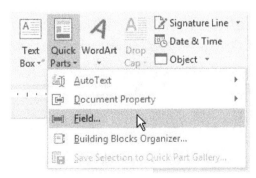

Step 4: In the Field dialog box, select the field you want to enter. You can select an option from the Categories drop down list to narrow down the options.

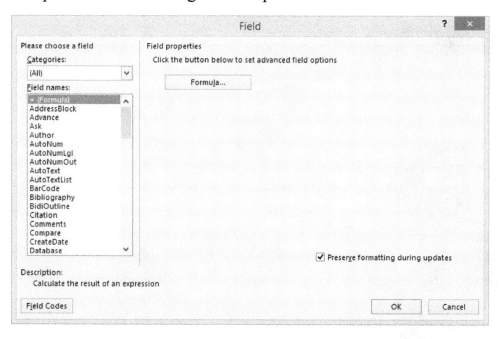

Step 5: Adjust the Field Properties and Field Options, depending on the field that you selected.

Step 6: Select OK.

To remove the field, just select it and delete the text.

Linking a Form to a Database

To insert data from a database into a form, use the following procedure.

Step 1: Open the Options dialog box by selecting Options from the Backstage View.

Step 2: Select the Customize Ribbon tab.

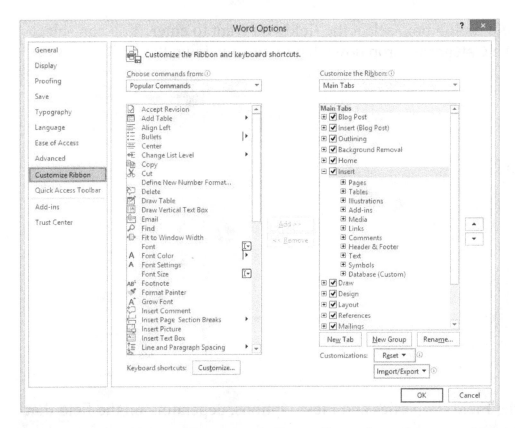

Step 3: In the Choose Commands From list, select Command Not in the Ribbon from the drop-down list.

Step 4: Highlight Insert Database from the list.

Step 5: On the Customize the Ribbon list, select the Custom Group where you want to include the command. See the previous module for information on creating a custom group.

Step 6: Select OK.

Now, with the Insert Database command available, select it in the location of your form where you want to include the database records.

Word displays the Database dialog box.

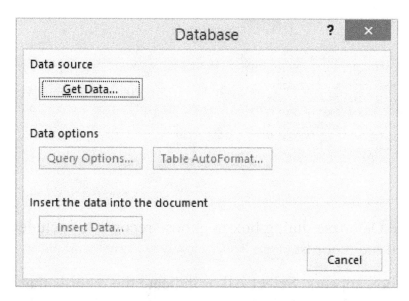

Step 1: Select Get Data.

Step 2: In the Select Data Source dialog box, navigate to the location of the database you want to use. Highlight it and select Open.

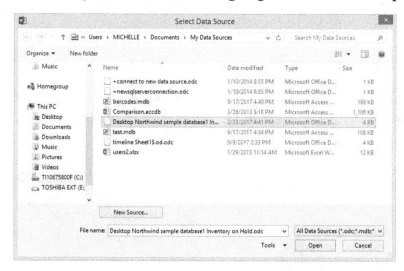

Step 3: If your database contains more than one table, the Select Table dialog box appears. Highlight the table you want to use and select OK.

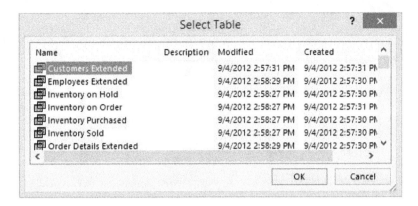

Step 4: Select Insert Data in the Database dialog box to choose records to include if you do not want to include all the records.

Step 5: The Insert Data dialog box appears. Select All or indicate the records that you want to include. You can check the Insert Data as Field checkbox if desired.

Add Links to Recent Files in Document

When you create a new document, you may want to include hyperlinks to other files you have worked on recently.

Add a link to a recently used file

Step 1: Place the cursor at the location in the file where you want to insert the link.

Step 2: Click the Insert ribbon.

Step 3: Click the drop-down arrow to the right of Link.

Step 4: Choose and click a file from the list. If you do not see the file you want to insert, click Insert Link at the bottom of the gallery and then navigate to the file you want to insert.

Ease of Access Options

New accessibility options are available in the Microsoft Office 2016 Suite of applications. You can set Feedback options, application display options, automatic alternative text, and document display options. Accessibility options help those with disabilities.

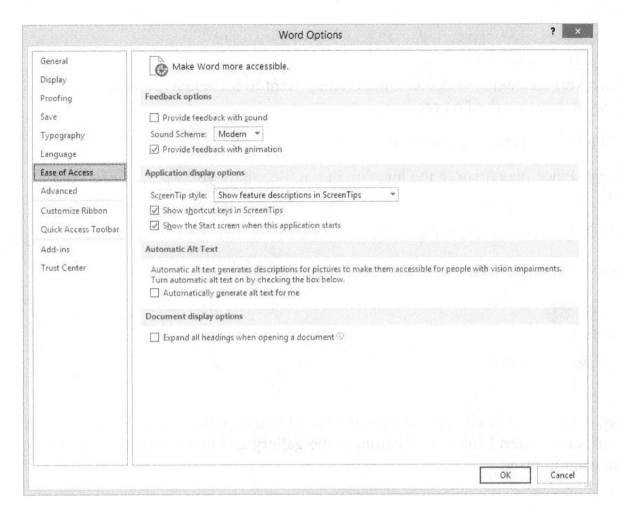

Icon Library

You can add visual impact to a document by inserting icons. You can choose for over 500 icons related to various topics.

Step 1: Click the Insert tab of the Ribbon.

Step 2: Click Icons in the Illustrations menu.

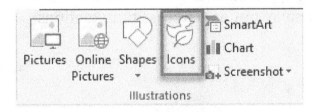

Step 3: Click the category you are looking for and then select the desired icon.

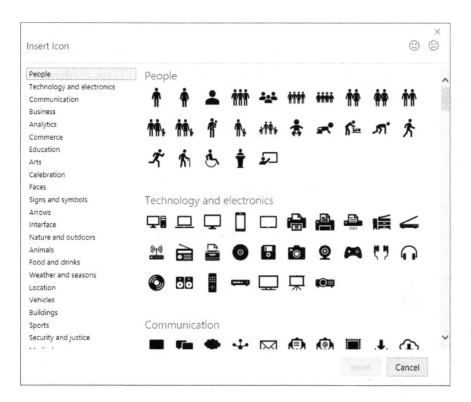

Step 4: Click Insert.

Improved Office Sounds

Sound effects can provide audio cues which can enhance productivity in Microsoft Word. A cue sound may tell you if the options on the screen change or can confirm an action has been completed. This improves accessibility in the applications.

Turn Sound Effects Off or On

Making changes to Provide Feedback with Sound in Word 2016 will change the settings in all the Microsoft Office programs.

Step 1: Click the File tab of the Ribbon.

Step 2: Click Options.

Step 3: Click the Ease of Access menu.

Step 4: Select or clear the checkbox next to Provide feedback with Sound". You can also adjust the sound theme from the Sound theme drop down menu.

New versions of office give you the option to draw freehand notations or shapes, and gives you the ability to highlight text on the new Draw ribbon. This ribbon includes the following tools:

- Eraser – to erase any items you add to the page.
- Pens used to write or highlight
- Ink to shape to draw a shape and then have the shape snap to the diagram and fill with the pen color
- Ink to Math to insert math equations into the document. It can also open the equation editor.

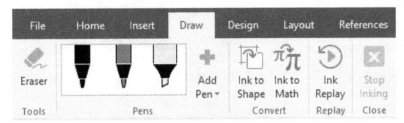

The new tools give you the option to draw with a digital pen, your finger, or a mouse. If your device is touch enabled, this ribbon is on by default. If the device is not touch enabled, you will need to turn on the ribbon in the Options menu.

Step 1: Go to the File ribbon and click Options.

Step 2: Select Customize Ribbon.

Step 3: Add a check mark next to the check box labeled Draw in the box on the right side of the dialog box.

Write, draw, or highlight text

The pen is portable and customizable. You define the pens they are then available in Microsoft Word, Microsoft Excel, and Microsoft PowerPoint.

Step 1: Tap a pen or highlighter on the **Draw** ribbon.

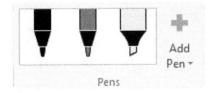

Step 2: Open the pen menu to set the Thickness and Color options for the pen. Select the preferred color and size.

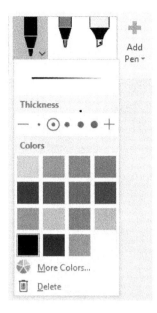

- Five predefined pen thicknesses from .25 mm to 3.5 mm. Select a thickness or use the minus or plus sign to make the pen thinner or thicker.
- Sixteen solid colors are available on the menu,
- Tap **More Colors** to see more options for colors

Step 3: Write or draw on the touch screen. Once you draw an ink shape, it behaves like a shape you are used to working with in Office. You can select the shape, move or copy it, change its color, and pivot its position.

Step 4: Select Stop Inking on the Draw tab to stop inking.

Erase ink

Step 1: Click the Draw ribbon select the Eraser in the Tools menu.

Eraser

Tools

Step 2: Drag the eraser over the ink you want to remove with your pen, mouse, or finger.

LaTeX Math Equation Syntax Added

Insert Built-in Equation

Step 1: Click the Insert tab on the Ribbon.

Step 2: Click the Equation expanded menu.

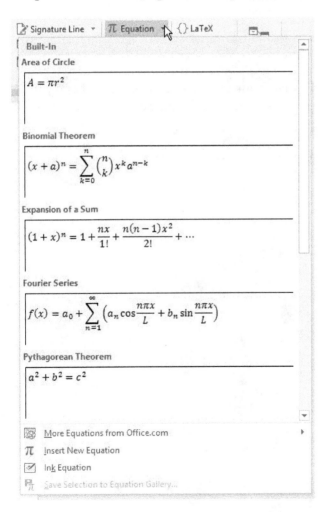

Step 3: Choose the equation you want to insert from the gallery. After the equation is inserted, the Design tab of the Equation Tools will be activated. This toolset has the symbols and structures that can be added to the equation.

When the Equation Tools ribbon is activated, you will be able to select LaTeX in the conversions menu.

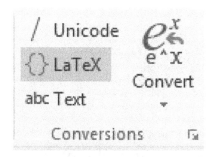

Ink to Math Convert

On touch and pen enabled devices you can write equations using a stylus, mouse, or your finger.

Step 1: Click the Draw ribbon.

Step 2: Click Ink to Math in the convert menu.

Step 3: Use a stylus, mouse, or your finger to write the equation.

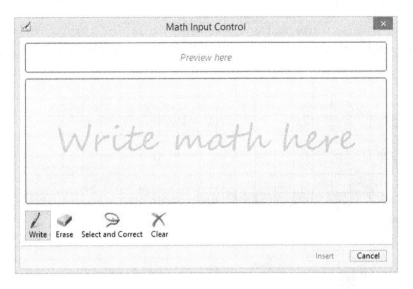

Step 4: Click Insert when you are done writing the equation.

Learning Tools

Learning Tools help you improve your ability to pronounce words correctly thereby boosting your reading skills. You do this by improving your ability to pronounce words correct (decoding), your ability to read quickly and accurately (fluency), and the ability to understand what you read (comprehension).

Learning Tools in Word 2016 provides the following tools:

Read Aloud - lets you hear your document in the default speech language of the computer, while simultaneously highlighting each word in the document.

Syllables Command – shows you breaks between syllables of words.

Text Spacing – increases the spacing between words, characters, and lines, improving scan ability of the content.

Column Width – is used to change the width of a line length.

Page color – can be set to sepia (print feel), inverse (white text on black background), or None

Access Learning Tools

Step 1: Click the View tab in the Ribbon.

Step 2: Click Learning Tools in the Immersive menu.

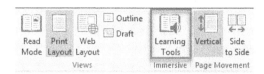

Step 3: Word 2016 will switch to the Web Layout view when the Learning Tools ribbon is opened.

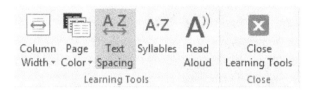

You can make the following changes using the Learning Tools ribbon.

Column width – changes the width of the line length to one of four options: very narrow, narrow, moderate, or wide.

Page color – Changes the page color to one of the following options: sepia (print feel – pale yellow page), inverse (white text with a black background), or none.

Text Spacing – changes the amount of space between letters, words, and paragraphs.

Syllables – shows the breaks between syllables in words in the content. In editing mode, you can make edits and the syllables will appear as you type.

Read Aloud – you can hear the document read aloud while each word is simultaneously highlighted. If you type while Read Aloud mode is engaged, the narration will pause while you make edits and then resume when the edits have been completed.

Read-Aloud

This feature allows the text to be read aloud while simultaneously highlighting the words in the document.

Step 1: Click the Review tab.

Step 2: Select Read Aloud in the Speech menu.

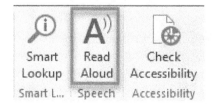

Use the following buttons that appear in the top right corner of the screen to control the reading.

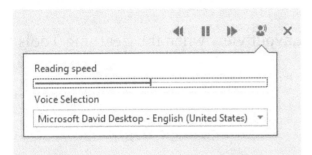

Play/Pause icon – Starts and Stops the narration.

Next - The narrator skips the current sentence and starts reading from the next paragraph.

Previous - The narrator skips the current sentence and start reading from the previous paragraph.

Settings menu - Change the reading speed and the voice

Close - Stops the narration and exits the Read Aloud mode

Narration starts reading from the position of the cursor in Word's editing views or the top of the page in Word's Read Mode by default. If you have selected a word, it will start reading from the selected word. Select a block of text before starting Read Aloud to will limit the narrator to only that selection.

Read Aloud Keyboard Shortcuts

To do this	Press
Start/Exit Read Aloud	CTRL+ALT+SPACE
Pause Read Aloud	CTRL+SPACE
Increase reading speed	ALT+RIGHT
Decrease reading speed	ALT+LEFT
Read previous paragraph	CTRL+RIGHT
Read next paragraph	CTRL+LEFT

Learning Tools in Read Mode can be used only while reading a document.

Step 1: Click the View tab from the ribbon.

Step 2: Click Read Mode in the Views group.

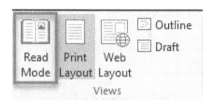

Or click Read Mode in the status bar.

Step 3: Do any or all the following from the View menu while in Read Mode.

View a Page Side by Side

If you would like to read through a document more like a book, instead of using the continuous scrolling.

Step 1: Click the View tab of the Ribbon.

Step 2: Click Side to Side in the Page Movement menu.

While looking at side by side view, a zoom menu will activate enabling you to look at the pages as thumbnails. Use thumbnails to quickly navigate through the document.

You also have shortcuts for getting to thumbnails which are available when the side by side option is selected. Use CTRL + the mouse wheel or on a touchscreen pinch to zoom out to thumbnails.

Tell Me

Instead of searching the online help or in Word, you can use the Tell Me feature to look for the solution you need. Click in the "tell me what you want to do" area and type your request.

To do this	Press
Insert endnote	ALT + CTRL + D
Insert footnote	ALT + CTRL + F
Switch to print preview	ALT + CTRL + I
Go to end of window	ALT + CTRL + Page Down
Go to top of window	ALT + CTRL + Page Up
Split document window	ALT + CTRL + S
Switch between last four places you have edited	ALT + CTRL + Z
Go back 1 page	ALT + Left Arrow
Go to "Tell me what you want to do" and Smart Lookup	ALT + Q
Insert comment (in Revision task pane)	ALT + R, C
Select Spelling & Grammar	ALT + R, S
Select Review tab on ribbon	ALT + R, then Down Arrow to move to commands on this tab.
Go forward 1 page	ALT + Right Arrow
Go to previous footnote	ALT + SHIFT + <
Go to next footnote	ALT + SHIFT + >

To do this	Press
Close Reviewing Pane if open	ALT + SHIFT + C
Remove document window split	ALT + SHIFT + C or ALT + CTRL + S
Mark a table of authorities' entry (citation)	ALT + SHIFT + I
Mark a table of contents entry	ALT + SHIFT + O
Mark an index entry	ALT + SHIFT + X
Move around preview page when zoomed in	Arrow keys
1 paragraph down	CTRL + Down Arrow
Move to last preview page when zoomed out	CTRL + End
Go to end of document	CTRL + End
Open search box in Navigation task pane	CTRL + F
Go to a page, bookmark, footnote, table, comment, graphic, or other location	CTRL + G
Replace text, specific formatting, and special items	CTRL + H
Move to first preview page when zoomed out	CTRL + Home
Go to beginning of document	CTRL + Home
Insert hyperlink	CTRL + K
Move 1 word to left	CTRL + Left Arrow
Create new document	CTRL + N
Open document	CTRL + O

To do this	Press
Print document	CTRL + P
Go to top of next page	CTRL + Page Down
Go to top of previous page	CTRL + Page Up
Move 1 word to right	CTRL + Right Arrow
Save document	CTRL + S
Turn change tracking on or off	CTRL + SHIFT + E
Move 1 paragraph up	CTRL + Up Arrow
Close document	CTRL + W
Move down 1 line	Down Arrow
Move to end of line	End
Refresh	F9
Move to beginning of line	Home
Move 1 character to left	Left Arrow
Down 1 screen (scrolling)	Page Down
Up 1 screen (scrolling)	Page Up
Move by 1 preview page when zoomed out	Page Up or Page Down
Move 1 character to right	Right Arrow
Go to a previous revision	SHIFT + F5
Move 1 cell to the left (in a table)	SHIFT + Tab

To do this	Press
Move 1 cell to the right (in a table)	Tab
Move up 1 line	Up Arrow

Edit and Move Text & Graphics

To do this	Press
Extend a selection to end of window.	ALT + CTRL + SHIFT + Page Down
Create new building block	ALT + F3
Copy header or footer used in previous section of a document	ALT + SHIFT + R
Delete 1 character to the left	Backspace
Select All	CTRL + A
Delete 1 word to the left	CTRL + Backspace
Copy selected text or graphics to Clipboard	CTRL + C
Delete 1 word to right	CTRL + Delete
Cut to the Spike (Spike is a feature that allows you to collect groups of text from different locations and paste them in another location)	CTRL + F3
Extend a selection to end of a paragraph	CTRL + SHIFT + Down Arrow
Extend a selection to end of a document	CTRL + SHIFT + End
Paste (Insert) Spike contents	CTRL + SHIFT + F3

To do this	Press
Select vertical block of text	CTRL + SHIFT + F8, and then use the arrow keys; press Esc to cancel selection mode
Extend selection to beginning of a document	CTRL + SHIFT + Home
Extend selection to beginning of a word	CTRL + SHIFT + Left Arrow
Extend selection to end of a word	CTRL + SHIFT + Right Arrow
Extend selection to beginning of a paragraph	CTRL + SHIFT + Up Arrow
Paste most recent addition or pasted item from Office Clipboard	CTRL + V
Cut selected text or graphics to Office Clipboard	CTRL + X
Undo last action	CTRL + Z
Delete 1 character to right	Delete
Turn extend mode off	Esc
Move text or graphics once	F2 (then move the cursor and press Enter)
Turn extend mode on	F8
Increase size of a selection	F8 (press once to select a word, twice to select a sentence, and so on)

To do this	Press
Select nearest character	F8, and then press Left Arrow or Right Arrow
Extend selection to a specific location in a document	F8+arrow keys; press Esc to cancel selection mode
Open Office Clipboard	Press ALT + H to move to the Home tab, and then press F,O.
Select Text	SHIFT + Arrow Keys
Extend selection 1 line down	SHIFT + Down Arrow
Extend selection to end of a line	SHIFT + End
When a building block is selected, display the associated shortcut menu	SHIFT + F10
Copy text or graphics once	SHIFT + F2 (then move the cursor and press Enter)
Reduce size of selection	SHIFT + F8
Extend selection to beginning of a line	SHIFT + Home
Extend a selection 1 character to left	SHIFT + Left Arrow
Extend a selection 1 screen down	SHIFT + Page Down
Extend a selection 1 screen up	SHIFT + Page Up
Extend a selection 1 character to right	SHIFT + Right Arrow
Extend selection 1 line up	SHIFT + Up Arrow

To do this	Press
Select entire table	ALT + 5 on the numeric keypad (with Num Lock off)
Move to last cell in row	ALT + End
Move to first cell in row	ALT + Home
Move to last cell in column	ALT + Page Down
Move to first cell in column	ALT + Page Up
Move 1 row down	ALT + SHIFT + Down Arrow
Move 1 row up	ALT + SHIFT + Up Arrow
Extend a selection (or block).	CTRL + SHIFT + F8, and then use the arrow keys; press Esc to cancel selection mode
Tab characters in a cell	CTRL + Tab
Move to next row	Down Arrow
Add new paragraphs in a cell	Enter
Extend selection to adjacent cells	Hold down Shift and press an arrow key repeatedly
Select preceding cell's contents or a previous cell in a row	SHIFT + Tab

To do this	Press
Select the next cell's contents or next cell in a row	Tab
Move to previous row	Up Arrow
	Use arrow keys to move to end of row, either first cell (leftmost) in row or to last cell (rightmost) in row.
	From first cell in row, press SHIFT + ALT + End to select row from left to right. From last cell in row, press SHIFT + ALT + Home to select row from right to left.
Select entire row	
	Use arrow keys to move to column's top or bottom cell, and then do one of the following:
	Press SHIFT + ALT + Page Down to select column from top to bottom. Press SHIFT + ALT + Page Up to select column from bottom to top.
Select a column	

To do this	Press
Apply Heading 1 style	ALT + CTRL + 1
Apply Heading 2 style.	ALT + CTRL + 2
Apply Heading 3 style.	ALT + CTRL + 3
Start AutoFormat	ALT + CTRL + K
Open Styles task pane	ALT + CTRL + SHIFT + S
Close Styles task pane	CTRL + Space Bar + Arrow Key to Select Close
Decrease font size by 1 point	CTRL + [
Increase font size by 1 point	CTRL +]
Add or remove 1 line space preceding a paragraph	CTRL + 0 (zero)
Single-space lines	CTRL + 1
Double-space lines	CTRL + 2
Set 1.5-line spacing	CTRL + 5
Apply Bold	CTRL + B
Open Font dialog box to change formatting of characters	CTRL + D
Toggle paragraph between centered and left-aligned	CTRL + E
Apply subscript formatting (automatic spacing)	CTRL + Equal Sign

To do this	Press
Apply italic formatting	CTRL + I
Switch a paragraph between justified and left-aligned	CTRL + J
Left align a text or paragraph	CTRL + L
Indent a paragraph from left	CTRL + M
Remove paragraph formatting	CTRL + Q
Switch a paragraph between right-aligned and left-aligned	CTRL + R
Display nonprinting characters	CTRL + SHIFT + * (asterisk on numeric keypad does not work)
Decrease font size	CTRL + SHIFT + <
Increase font size	CTRL + SHIFT + >
Format all letters as capitals	CTRL + SHIFT + A
Copy formatting	CTRL + SHIFT + C
Double-underline text	CTRL + SHIFT + D
Open Font dialog box to change font	CTRL + SHIFT + F
Apply hidden text formatting	CTRL + SHIFT + H
Format letters as small capitals	CTRL + SHIFT + K
Remove a paragraph indent from left	CTRL + SHIFT + M
Apply Normal style	CTRL + SHIFT + N

To do this	Press
Apply superscript formatting (automatic spacing)	CTRL + SHIFT + Plus Sign
Change selection to Symbol font	CTRL + SHIFT + Q
Open Apply Styles task pane	CTRL + SHIFT + S
Reduce hanging indent	CTRL + SHIFT + T
Paste copied formatting	CTRL + SHIFT + V
Underline words but not spaces	CTRL + SHIFT + W
Remove manual character formatting	CTRL + Spacebar
Create hanging indent	CTRL + T
Apply an underline	CTRL + U
Select Styles task pane	F6
Review text formatting	SHIFT + F1 (then click the text with the formatting you want to review)
Change case of letters	SHIFT + F3

Function Key Reference

To do this	Press
Go to next field	ALT + F1
Display Selection task pane.	ALT + F10

To do this	Press
Display Microsoft Visual Basic code	ALT + F11
Exit Word	ALT + F4
Restore program window size	ALT + F5
Move from an open dialog box back to document, for dialog boxes that support this behavior	ALT + F6
Find next misspelling or grammatical error	ALT + F7
Run macro	ALT + F8
Go to previous field	ALT + SHIFT + F1
Display menu or message for available action	ALT + SHIFT + F10
Choose Table of Contents button in Table of Contents container when container is active	ALT + SHIFT + F12
Choose Save command	ALT + SHIFT + F2
Display Microsoft System Information	CTRL + ALT + F1
Choose the Open command	CTRL + ALT + F2
Maximize document window	CTRL + F10
Choose Open command	CTRL + F12
Choose Print Preview command	CTRL + F2
Close window	CTRL + F4
Go to next window	CTRL + F6
Choose Print command	CTRL + SHIFT + F12

To do this	Press
Edit bookmark	CTRL + SHIFT + F5
Go to previous window	CTRL + SHIFT + F6
Extend a selection or block	CTRL + SHIFT + F8, and then press an arrow key
Get Help or visit Office.com	F1
Show Key Tips	F10
Choose Save As command	F12
Move text or graphics	F2
Repeat last action	F4
Choose Go To command (Home tab)	F5
Choose Spelling command (Review tab)	F7
Start context-sensitive Help or reveal formatting	SHIFT + F1
Choose Save command	SHIFT + F12
Copy text	SHIFT + F2
Repeat Find or Go To action	SHIFT + F4
Go to previous pane or frame (after pressing F6)	SHIFT + F6
Choose Thesaurus command (Review tab, Proofing group)	SHIFT + F7

To do this	Press
Open Insert an object dialog box	Alt + N, J, J
Select object and create	Down Arrow + Enter
Insert or browse to object in a file	CTRL+TAB, TAB, File Name
Edit object	(1) Cursor Right of Object + SHIFT + Right Arrow (2) Shift + F10 (3) Tab, Enter, Enter
Insert SmartArt graphics	(1) ATL + N + M (2) Arrow keys to select graphic type (3) Tab, Arrow keys to select graphic (4) Enter
Insert WordArt	(1) Alt + N + W to select WordArt (2) Type text (3) Esc to select, arrow keys to move object (4) Esc to return to document

To do this	Press
Insert copyright symbol	ALT + CTRL + C
Insert an em dash	ALT + CTRL + Minus Sign (on the numeric keypad)
Insert an ellipsis	ALT + CTRL + Period
Insert registered trademark symbol	ALT + CTRL + R
Insert the trademark symbol	ALT + CTRL + T
Insert ANSI character for specified ANSI (decimal) character code. For example, to insert the euro currency symbol, hold down Alt and press 0128 on the numeric keypad.	ALT + the character code (on the numeric keypad)
Find out Unicode character code for selected character	ALT + X
Insert single closing quotation mark	CTRL + ' (single quotation mark), ' (single quotation mark)
Insert double closing quotation marks	CTRL + ' (single quotation mark), SHIFT + ' (single quotation mark)
Insert double opening quotation marks	CTRL + ` (single quotation mark), SHIFT + ' (single quotation mark)

To do this	Press
Insert single opening quotation mark	CTRL + ` (single quotation mark)
Insert page break	CTRL + Enter
Insert field	CTRL + F9
Insert optional hyphen	CTRL + Hyphen
Insert an en dash	CTRL + Minus Sign (on the numeric keypad)
Insert column break	CTRL + SHIFT + Enter
Insert nonbreaking hyphen	CTRL + SHIFT + Hyphen
Insert nonbreaking space	CTRL + SHIFT + Spacebar
Insert AutoText entry	Enter (after you type the first few characters of the AutoText entry name and when the ScreenTip appears)
Insert line break	SHIFT + Enter
Insert Unicode character for the specified Unicode (hexadecimal) character code.	The character code, ALT + X

To do this	Press
Edit mail-merge data document	ALT + SHIFT + E
Insert merge field	ALT + SHIFT + F
Preview mail merge	ALT + SHIFT + K
Print merged document	ALT + SHIFT + M
Merge document	ALT + SHIFT + N

Move Around Ribbon

To do this	Press
Select active tab of ribbon and activate access keys	Alt or F10. Use access keys or arrow keys to move to a different tab.
Open File ribbon to use Backstage view	ALT + F
Open Design tab to use themes, colors, and effects	ALT + G
Open Home tab to use common formatting commands, paragraph styles, or the Find tool	ALT + H
Open Mailings tab to manage Mail Merge tasks or to work with envelopes and labels	ALT + M
Open Insert tab to insert tables, pictures and shapes, headers, or text boxes	ALT + N
Open Layout tab to work with page margins, page orientation, indentation, and spacing	ALT + P

To do this	Press
Open "Tell Me" box on ribbon to type a search term for Help content	ALT + Q, then enter the search term
Open Review tab to use Spell Check, set proofing languages, or to track and review changes to document	ALT + R
Open References tab to add a table of contents, footnotes, or a table of citations	ALT + S
Open View tab to choose a document view or mode, such as Read Mode or Outline view	ALT + W
Expand or collapse ribbon	CTRL + F1
Move down, up, left, or right among items on ribbon	Down Arrow, Up Arrow, Left Arrow, or Right Arrow
Finish modifying a value in a control on ribbon, and move focus back to document	Enter
Move focus to next pane	F6
Display shortcut menu for selected item	SHIFT + F10
Activate selected command or control on ribbon	Spacebar or Enter
Move focus Forward to commands on ribbon	Tab
Move focus backward to each command on ribbon	SHIFT + Tab

To do this	Press
Switch to Draft view	ALT + CTRL + N
Switch to Outline view	ALT + CTRL + O
Switch to Print Layout view	ALT + CTRL + P
Show all headings with Heading 1 style	ALT + SHIFT + 1
Expand or collapse all text or headings	ALT + SHIFT + A
Move selected paragraphs down	ALT + SHIFT + Down Arrow
Show first line of text or all text	ALT + SHIFT + L
Promote a paragraph	ALT + SHIFT + Left Arrow
Collapse text under a heading	ALT + SHIFT + Minus Sign
Show all headings up to Heading n	ALT + SHIFT + n
Expand text under a heading	ALT + SHIFT + Plus Sign
Demote a paragraph	ALT + SHIFT + Right Arrow
Move selected paragraphs up	ALT + SHIFT + Up Arrow
Switch to Read Mode view	ALT + W, F
Demote to body text	CTRL + SHIFT + N

To do this	Press
Insert a tab character	CTRL + Tab
Go to end of document	End

Work with Fields

To do this	Press
Insert a LISTNUM field	ALT + CTRL + L
Switch between all field codes and their results	ALT + F9
Insert a DATE field	ALT + SHIFT + D
Run GOTOBUTTON or MACROBUTTON from field that displays field results	ALT + SHIFT + F9
Insert a Page field	ALT + SHIFT + P
Insert a TIME field	ALT + SHIFT + T
Lock field	CTRL + F11
Insert empty field	CTRL + F9
Unlock field.	CTRL + SHIFT + F11
Update linked information in a Microsoft Word source document	CTRL + SHIFT + F7
Unlink field	CTRL + SHIFT + F9
Go to next field	F11
Update selected fields	F9

To do this	Press
Go to previous field	SHIFT + F11
Switch between a selected field code and its result	SHIFT + F9

Work with Language Bar

To do this	Press
Turn Japanese Input Method Editor (IME) on 101 keyboard on or off	ALT + ~
Set default languages	ALT + R, L
Open Set Proofing Language dialog box	ALT + R, U, L
Turn Chinese Input Method Editor (IME) on 101 keyboard on or off	CTRL + Spacebar
Review list of proofing languages	Down Arrow
Turn Korean Input Method Editor (IME) on 101 keyboard on or off	Right Alt